Ingenieurwissen

Plastics for external car components

Published by Verein Deutscher Ingenieure
VDI-Gesellschaft Kunststofftechnik

English translation by M. S. Welling
German translation by H. Domininghaus

VDI-Verlag GmbH
Verlag des Vereins Deutscher Ingenieure · Düsseldorf

CIP-Kurztitelaufnahme der Deutschen Bibliothek

Plastics for external car components / Hrsg.: Verein Dt. Ingenieure, VDI-Ges. Kunststofftechnik. – Düsseldorf: VDI-Verlag, 1980.
(Ingenieurwissen)
Dt. Ausg. u. d. T.: Kunststoffe für Außenteile im Automobilbau.
ISBN 3-18-404059-3

NE: Gesellschaft Kunststofftechnik

Printed in Germany

ISBN 3-18-404059-3

Preface

The changed situation in the energy and raw material sectors, coupled with the growing desire for comfort and simpler operation, as well as new safety and statutory regulations, have forced the car industry to consider the use of a wider range of materials. Most of the plastics being used at the moment are to be found inside the vehicle. Future growth rates in the high polymer field will, however, be concentrated on those plastics materials which will find use in the coachwork and external car components. This will lead to further reductions in the weight of cars and contribute towards reducing the number of parts subject to corrosion.

How popular the front and rear ends made of polyurethane, GRP or thermoplastic materials, used in the USA, will become in Europe is open to discussion. In this connection, the task of making work more pleasant for assembly line workers and the related change-over from assembly lines to separate assembly groups is also to be considered.

This book is also intended to explain the background for the somewhat slow rate of development in the use of plastics for external car components. The time has come to consider new application possibilities and put them into practice. Plastics could replace metals to a far greater extent and there is a definite and growing trend towards the use of these materials in vehicle construction.

Mannheim, March 1980

Hermann Halblitzel
Chairman of the Technical Committee "Plastics in Vehicles" of the VDI-Gesellschaft Kunststofftechnik

Contents

Prospects for plastics in external car components

Nuccio Bertone

In his daily work of creating new models, the designer generally assumes that his proposed model can actually be made from a given material. In correctly interpreting the given subject, innumerable factors have to be taken into account without interfering too much with the liberty of the designer regarding the choise of material.

It es expedient to allow the stylist great freedom so that he can express his creative ideas with a minimum of hindrance. This is the only way to enable him to introduce original projects. The designer should work as closely as possible with the stylist and seek to realize the end product using materials which best underline the formal quality and aesthetic aspects of the model. The choice of material is therefore always the job of the designer, who must take into consideration such factors as capital outlay, technologies, costs, weight etc.

The problem of the designer requires some consideration. The use of plastics in motor vehicles is usually regarded by the end user as an attempt on the part of manufacturers to save money. The customer's initial reaction to plastics is therefore negative. Consciously or subconsciously, the customer identifies a plastics product with „cheapness'' since, in his daily life, he is surrounded by cheap plastics objects. This creates in him a feeling of aversion, frequently also associated with the problem of environmental pollution. There is, furthermore, still the inclination to make cost comparisons between plastics and traditional materials when it is usually found that the latter cost slightly more.

The designer, whose aim it is to interpret aesthetic and functional problems, accepts with reluctance the neglected appearance of the plastics nowadays being used for external car components and feels that these do not guarantee any formal or aesthetic advantages over traditional materials.

There are, however, exceptions, such as the front and rear ends of the Porsche 928, figs. 1 and 2, which are perfect. These are functional, very well made and remarkebly easily assembled. In this instance, the designer

Fig. 1. Porsche 928, rear view

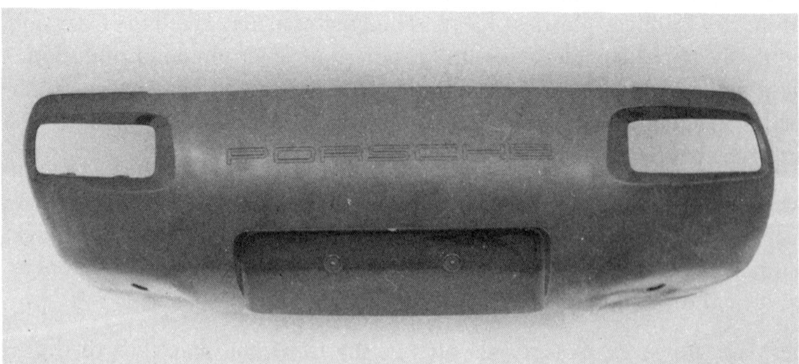
Fig. 2. Porsche 928, rear end

was able to design the front end and the rear view mirror in a pure and more original from than would have been possible with traditional materials. The high price of this car and the good image of Porsche have made it possible to use „quality" plastics in this case, these being perfectly acceptable to customers.

The purpose of the remarks is to provoke – to put research workers under pressure to develop, as soon as possible, inexpensive plastics materials which please the customer and posses the same properties as traditional materials as well as even more reliable and with better all-round properties. This is the only way of improving the image of plastics – an image which is still only „modest" – and to cause manufacturers and designers to turn away from traditional materials and replace them by plastics.

Present situation

A glance at present-day world production of cars made of plastics does not present a very encouraging picture. It is, of course, true to say that nearly all cars made today contain individual parts or assemblies made from plastics, but the number of models with an all-plastics body is only a few ten thousand compared with an annual world production of more than 30 million vehicles. One must also remember that, with but a few exceptions such as the GM Corvette, fig. 3, the Citroën Méhari, fig. 4, the Matra Bagheera, fig. 5, the Reliant Scimitar, fig. 6 and the Lotus Esprit, fig. 7, the other models with plastics bodies are purely hand-made replicas, fig. 8, which do not have any of the features which inspire confidence and could guarantee an international market for the car.

Fig. 3. GM Corvette

Fig. 4. Citroën Méhari

Fig. 5. Matra Bagheera

Fig. 6. Reliant Scimitar

Fig. 7. Lotus Esprit

Fig. 8. BMW replica model

Fig. 9. Bertone Panther

Our own attempts to produce plastics cars (1967 Panther and 1976 Navajo, figs. 9 and 10) were unsuccessful compared with similar prototypes made of metal, mainly because they took longer to produce and GRP did not really give satisfactory results.

Fig. 10. Bertone Navajo

Things turned out rather better in the production of the Lancia „Stratos", fig. 11 and the FIAT Abarth 131 Rally, fig. 12. In both cases however, the numbers produced were only 500 per model. The basic conditions were light weight, low capital outlay and, most importantly, fast production to enable these models to be approved as quickly as possible as sports cars. From this point of view, the end result was achieved in full. The Stratos, although partly forgotten today, is still unsurpassed as a rally car. It is the

Fig. 11. Lancia Stratos

Fig. 12. Fiat Abarth 131 Rally

only which has won the rally world championship for three years running (1974, 75 and 76).

It has often been suggested to designers, without much success, to use plastics for parts to be integrated in our production models. A typical example is the Audi 50. One model, introduced to Audi in 1972, provided for a front end and a rear end partly made of plastic. This solution was not accepted at the time because it appeared premature amd altogether too original. There were also doubts about costs. Nowadays, this solution has been accepted in a number of lower and medium price range vehicles.

In another of our prototypes, the „Trapeze" built in 1973, fig. 13, the application of a strap round the car was suggested, this being well integrated with the bumpers. It was aesthetically pleasing and at the same time carried out a protective function. Various plastics components are mounted on the

Fig. 13. Trapeze

Fig. 14. Fiat X 1/9

body of the FIAT X 1/9, fig. 14, including a combined front end/spoiler, bumper, detachable roof made of SMC and others. In the prototype for the Ascot of 1977, fig. 15, the external parts, covering for column A, rain gutter etc. have been designed in a novel way. In the plastics model, these parts are coloured grey so that final assembly operations are simplified and quicker to carry out. Fixed properly to the body, these would neatly cover difficult parts of the car and help to reduce costs. Thanks to their flexibility and relative softness they would also contribute towards passive safety.

Fig. 15. Ascot

Impact with the columns A would then no longer be so critical. A similar solution would be feasible for the passenger compartment, safety being combined with style.

Future prospects

The table prepared by the Battelle Institute, fig. 16, shows how the weight of plastics used in vehicles has increased from 1 % in 1965 to 4 % in 1974 and almost 6 % in 1980. It may be assumed that this trend will continue to increase in future, development being even more rapid and intense. Weight is one of the most important factors to be considered in efforts to reduce fuel consumption. Work has alrealy started in this direction, for example in the United States, where legislation has been introduced which lays down that, by 1985, cars must have an average fuel consumption of 27.5 miles per gallon. As the table shows, there has been a definite tendency for American cars to be lighter since 1971, so that by 1980, the reduction in weight will be up to 20 % per car, compared with 1971. For 1985, General Motors are planning to meet the 27.5 miles per gallon consumption for their medium class vehicles, despite the fact that these will still weight 3 200 lbs.

The ratio of fuel consumption and weight depends on the overall concept of the vehicle in which new solutions – lighter materials and lightweight contruction – are introduced and influence the realization of a project in the creation of new production facilities. The more consistently this concept is carries through, the more fuel will be saved per kg of vehicle weight. Furthermore, the result depends on whether one decides to replace one part of the vehicle after another, or wether the car is to be newly constructed right from the start.

The general situation in Europe ist evidently quite different to that in the USA. The weight of the average new car in the USA (about 2 tons) is today still almost twice the average weight of a European car (about 1 ton). The average American car still does less than 12 miles per gallon, whilst the prescribed limit of 27.5 miles per gallon has already been achieved by European cars.

American car manufacturers are trying hard to produce smaller cars whose size is more like that of European cars. Nevertheless, these are still going to be bigger than those seen on European roads now and in the years to come.

This privileged position of the European car industry is, however, more apparent than real. Even though there are no laws for Europe as regards

Material used in a car, according to Battelle Institute

Material	1965		1968		1971		1974		1980	
	lbs/unit*	%	lbs/unit	%	lbs/unit	%	lbs/unit	%	lbs/unit	%
Cast iron	414	12,5	440	12,6	488	12,2	467	12,2	378	11,8
Carbon steels	1901	55,1	1977	54,3	2152	53,8	2030	53,0	1644	51,3
Special purpose steels	287	7,8	293	7,7	332	8,3	334	8,9	282	8,8
Aluminium	62	1,8	66	1,8	76	1,9	83	2,2	115	3,6
Other non-ferrous metals	129	3,7	124	3,6	132	3,3	93	2,4	61	1,9
Glas	97	2,8	98	2,7	104	2,6	96	2,5	80	2,5
Plastics	35	1,0	73	2,0	104	2,6	152	4,0	190	5,9
Rubber	166	4,8	182	5,0	204	5,1	199	5,2	160	5,0
Miscellaneous**	359	10,5	375	10,3	408	10,2	378	9,6	295	9,2
Total	3450	100	3628	100	4000	100	3841	100	3205	100

* 1 lb = 0.4536 kg

** adhesives, lubricants, cotton, welded seams

Fig. 16. Weight of different materials in a car, 1976 – 1980 (Battelle Institute)

Fig. 17.
Mercedes Economobile

fuel consumption in 1985, it is likely that more restrictions will be introduced since we are totally dependent upon the oil-producing countries. It is not mere chance that firms like Mercedes-Benz are engaged on projects which only apperently are hobbies. The „Economobile Mercedes", fig. 17, is a typical example. Weighing only 50 kg, it recently set a world record for a distance of 1284.13 km, during which it used only one litre of fuel per 100 km.

It is not illogical to predict that it may well be necessary for the European car industry to build a car in the nineties which will offer comfortable accomodation for 5 people and have a guaranteed fuel consumption of no more than 50 miles per gallon.

The American car industry is today being put under pressure in two respects:

the laws for 1985

considerable weight reduction

To save weight, the Americans have not chosen the path towards a global solution but have concentrated on individual components which can quickly and easily be replaced by lighter ones. A typical example is the replacement of metals for the front parts of American cars with the so-called „soft nose". This not only reduces the weight but also saves a number of assembly operations as well energy during the manufacturing process. The current process of substituting components enables inventors, engineers and designers to accumulate experience which is very important in finding the right path towards weight reduction, i. e. the path towards complete re-design.

Initially, perhaps, this substitution process will find increased possiblities for using structural and functional plastics components rather than external car components, mainly for stylistic rather than logical, functional reasons. The reliability of certain plastics components, whose permitted loading is exactly defined, is better than for steel components and one should therefore try to persuade the customer to accept assemblies such as seats, wheels, leaf springs, drive shafts and other mechanical parts made of plastic rather than certain external carbody components where the customer requires not only reliability but also an attractive appearance.

The different ageing characteristics of painted surfaces, depending on whether the paint has been applied to metal or plastic, means that the car owner finds he has a two-tone car after a few years. A painted metal surface will have a good finish everywhere, something that is very difficult to achieve with plastic parts. Many customers, especially those of the more expensive models, find the surface finish of plastics components unacceptable, e. g. the Ferrari 308 GTB, fig. 18. This was originally introduced with a very well made GRP body, but nevertheless the company found itself forced, after only a few units, to return to the use of steelplate since it had proved

Fig. 18. Ferrari 308 GTB

Fig. 19.
Europa 3 rocket

Fig. 20. Boeing 767 (model)

impossible to get customers to accept even a few tiny faults which had not been present in metal bodies.

Past use of „cheap" plastics which, nevertheless, had been expensive to purchase, probably has a detracting effect on cars. In aircraft construction on the other hand, where there are other economic factors to be considered in the selection of materials, plastics are experiencing a real boom. Air Italia for example, in connection with the ELDO research programme, constructed the protective shields for the Europa 3 rocket, fig. 19, entirely from plastics. The same applies to the new Boeing 767, fig. 20, where the side rudders and height rudder are made entirely of plastic. These are components measuring 6 m and 12 m respectively and have withstood greater loads than similar parts made from aluminium alloy whilst, at the same time, being 20 % lighter in weight.

Naturally, the parts are made from carbon fibre reinforced plastics, which are still too expensive for use in cars, apart from the formula 1 cars, e. g. the new Ferrari T5, fig. 21, whose „miniskirts" and other carbody components are made of carbon fibre reinforced plastic whose price is kept so high by a vicious circle:

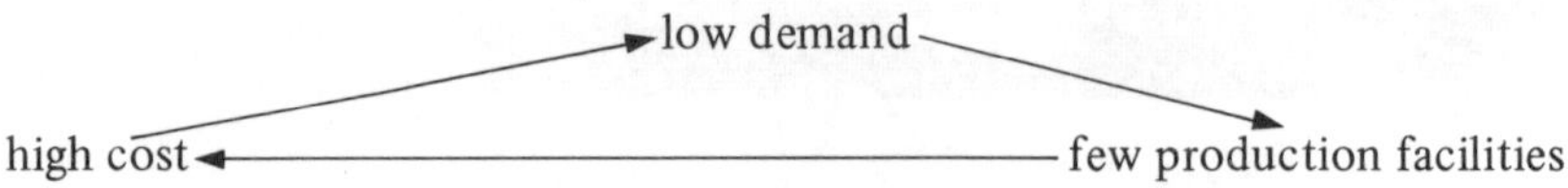

Fig. 21. Ferrari T 5

The use of larger quantities would result in a dramatic price reduction. This trend is already becoming noticeable and the materials are starting to be used in branches other than the aircraft and rocket industries, and ranging from cars to sports articles. This trend could expand in future if the energy necessary to produce these materials could be available more cheaply through the expansion of nuclear power stations. The same argements also apply to the new product „KEVLAR" – a very high impact fibre made by DuPont, whose use in cars is, however, unthinkable at present on account of its very high cost.

Apart from the problem of profitability of plastics, there is still considerable reluctance, even today, for the industry to change its structure and invest in the required new technologies. The hundred years during which steelplate has been used cannot be erased in a few years. The techniques of welding, painting etc. have greatly influenced the design of a steelplate body and brought it to limits which can hardly be improved upon. These criteria must be revised if one is to consider using plastics. One need only think of painting. Because of its anti-corrosive function, paint froms a very important part of car construction and represents one of the most expensive investments. Some of these production stages would no longer be necessary when using plastics, so that cost would certainly be lowered.

Engineers as well as workmen would have to get used to these new materials gradually, get to know their properties and gain the experience and familiarity which are absolutely essential for the large-scale use of plastics. The same applies to the customer service network and carbody workshops which have specialised in the repair of metal bodies and are unfamiliar with plastics repairs.

Fig. 22. Ford project, graphite fibre reinforced plastic vehicle

It is fairly certain that the present policy of replacing „components by components" is the right one. In this connection we should mention the new Ford project GFRPV (graphite fibre reinforced plastic vehicle), fig. 22, which is the first serious experiment involving the total replacement of the metal body by one made of plastic. This is the logical evolution which one day could lead to total redesign.

Through such experiments, the designer will find it possible to express himself with new shapes and new ideas. The new material will enable him to „dare" produce a new shape which would be totally impracticable, for technical reasons, with traditional materials.

In future one will have to offer the customer a car whose quality will have to be at least as good as that of a metal car, e. g. long-lasting good appearance, colour, finish, ease of repair etc., if possible at the same price or even cheaper. Once this has become possible, the customer will refuse a metal car in favour of one made of plastic.

New pressures and the alternative: plastics

Hermann Hablitzel and *Klaus Dieter Johnke*

The theme of this year's annual conference – „Plastics for external car components" – was not chosen only because it seemed an appropriate complement to last year's theme „Plastics for the interior trim of vehicles" but also to facilitate public discussion of the many absurdities that have been raised by the press and by talks between raw material producers and the manufacturers of plastics car components. In such conversations, the term „lightweight construction" is very often mentioned in connection with high polymers. Plastics, however, can offer much more important properties than that of being lighter than other materials, especially when one considers the future. There are other reasons why plastics are being used increasingly in vehicle building.

Introduction

Plastics primarily serve *technical progress.* In certain areas they increase serviceability, improve comfort and meet the regulations designed to increase vehicle safety by preventing accidents and reducing the extent of accident damage. The careful use of plastics furthermore contributes towards reducing the noise level both inside and outside the vehicle. Naturally, some of the considerations are concerned with *lightweight construction and improvement of the* c_W *value.* Both should be regarded in conjunction with fuel saving during use and energy saving during vehicle manufacture. There has been a great deal of activity in this field but the desired success has not yet been achieved, unfortunately. Improving the c_W value by using plastics implies that requirements which would serve to reduce the c_W value can only be met technically and economically by using plastics. Here we should mention front parts, spoilers, wind deflectors, coverings underneath the engine compartment and underbody sealants etc.

Progress in production techniques is another reason for examining the use of plastics in vehicle construction. Great efforts are being made in all areas of production engineering to make work easier – humanisation at the place of work. The large numbers involved in building a car, and the extremely repetitive nature of the assembly process make it necessary to simplify overhead assembly operations and heavy work on the assembly line and to effect improvements by means of new equipment. The assembly process should be transferred from narrow confines of the conveyor belt to individual working areas. In this respect, special importance is attached to the unit system of construction, i. e. units considered and fitted collectively. The consistent use of such a system helps the manufacturing process – and thus also the end product supply – to be more flexible. This is a very important consideration, bearing in mind the many different designs required by customers. Last but not least, advantages are also likely to result for after-sales service.

The various points we have mentioned overlap considerably, but as the following examples demonstrate, plastics offer important advantages over conventional materials if they are used for applications that make full use of their special properties. In vehicle building especially, there are clear limits insofar as plastics are concerned, which should be observed, e. g. as regards long-term performance, performance at high and low temperatures and when subject to sudden deformation. There has, however, been progress in recent years in these fields, too, so that ,,tailor-made plastics" are today available for a wide range of applications.

Technical progress through plastics

Plastics as a replacement material

When one speaks of the use of plastics in vehicles, emphasis today is still on the interior trim for no less than two-thirds of all plastics used in vehicles replace materials such as felt, insulating materials, fabrics and claddings and only one-third replaces metal, fig. 1.

Since, however, the greater part of plastics materials is being used at present in the interior trim, future percentage increase rates for plastics will concern those materials used for car bodies and external components. In this area, plastics have assumed protective functions, e. g. bumper covers, ram strips, mudguard extensions, covers in wheel housings etc.

All-round protection through front, rear and side areas

Improved all-round protection is one of the most important objectives. VW, in 1976, introduced a new generation of plastics bumpers which is now being used in practically all the models made by the company, fig. 2. This

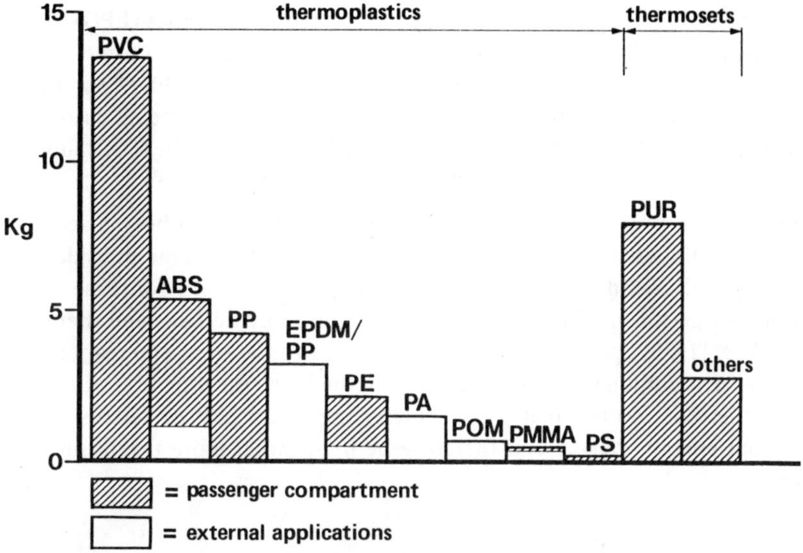

Fig. 1. Plastics applications in the VW Golf.

Fig. 2. Part of the bumper unit of the VW Golf with plastics covering (EPDM/PP).

design, which consists of a steel support covered with plastics (EPDM/PP), ensures that vehicles are given better protection. In the meantime, other designs have been developed, based on thermoplastics (modified polypropylene) with and without support. Other ideas, such as are being used by Mercedes (GRP supports covered with polyurethane), Porsche (GRP supports without covers) or Peugeot (PU bumper) are feasible.

In special models the unnecessary chrome trim has often been dispensed with and the use of impact resistant plastics at the front, rear and sides has achieved all-round protection. The most important plastics for these applications are glass fibre reinforced polyester resins, polyurethane and thermoplastics, fig. 3. Special importance is attached to the front and rear ends. These are components positioned at right angles to the direction of travel and therefore do not play a particularly important part in destroying impact energy. Plastics can therefore be used rather more readily in this case.

Fig. 3. All-round protection for the special model Renault 5 Turbo.

These new materials, specially modified to absorb impact energy, are already showing great promise and can be compared with steel components.

Improved comfort through use of plastics

Plastics are frequently being used where comfort is to be improved. This tendency is not so marked externally as in the passenger compartment. Nevertheless, one must remember that plastics afford better protection against heat and cold, so that the passenger will be surrounded by an even temperature. Moreover, sound transmission is less and very complex components used to reduce engine noise are more easily made from plastics than steel. Another point is that plastics components are more pleasant to the touch at very high and very low temperatures.

Legal requirements

To observe safety regulations it is, in many cases, possible only to use plastics components. Here we shall primarily deal with the reduced fuel consumption due to lighter weigth for everything we have discusses so far, led to increased weight. Covering bumpers with plastic increases the weight of VW vehicles by an average of about 4 kg. Although such bumpers afford very good protection against low-speed impact – also in impact with pedetrians – they replace the chrome of present-day bumpers and are also much bigger so as to afford greater protection. If one were to enlarge steel bumpers to give the same degree of protection, at least at the points of contact,as plastics bumpers, they would be even heavier. Maximum advantage has therefore been gained with the minimum weight.

There are, however, other places where the use of plastics brings greater advantages, e. g. in spoilers at the front and rear which, by reducing the air resistance factor (c_W value) reduces fuel consumption. Rain channel coverings used by Mercedes to improve the vehicle's airflow, or covers near the base, can likewise result in a reduction of the c_W value.

Reduction in c_W value and lightweight construction help to conserve energy

Reduction of the c_W value and lightweight construction would not have become possible without the use of plastics. Although every group of raw material manufacturers produces fictitious reasons for their supposed supremacy over others, an objective assessment should be possible through comparing relevant figures, thereby obtaining a more balanced picture of the true situation. Under no circumstances should the necessity to reduce fuel consumption be taken as justification for making cars more expensive.

In considering the economic points of view, it becomes apparent that material selection is very important, since 53 % of the production costs for a vehicle are for materials, fig. 4. Material selection is therefore in accordance with the following criteria:

technical progress

minimising costs

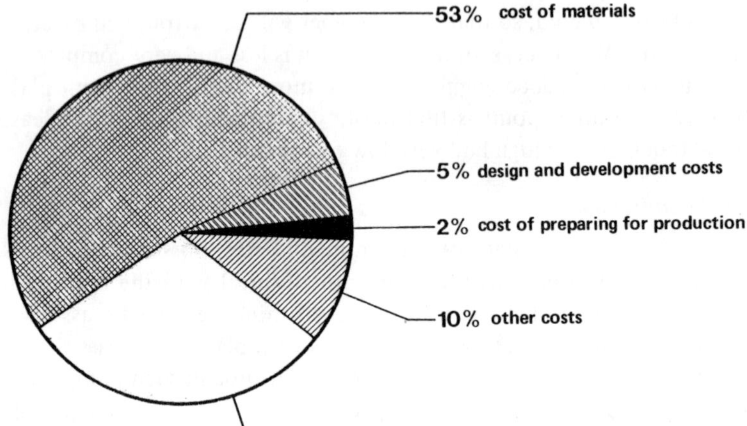

Fig. 4. Cost breakdown for the production of a new vehicle.

component optimisation

optimisation of production techniques.

Besause of the large amount of energy required to make raw materials and for their conversion into finished and semi-finished products, future emphasis will be on chosing energy-saving material production processes. Finished components with an unlimited life are unthinkable. Materials used to make mass produced articles should therefore be those which can be recycled. Future developments should therefore increasingly take into account not only functional component optimisation but also the long-term performance of components, the possibility of recycling and the use of energy-saving raw material manufacture and production techniques. Waste elimination is also not without importance.

Lightweight construction

All the above mentioned points of view habe to be considered when one is thinking about ,,lightweight construction". Greater attention should therefore be paid in future to the following points, which are all to be regarded in connection with the saving of energy and raw material:

lightweight construction with fullest use being made of material properties

improvement of long-term performance (reliability in use and maintenance of material worth)

energy-saving raw material manufacture
energy-saving production techniques
energy savings in operation
recycling
reduction of waste problems.

The problems of energy conservation and protection of raw material reserves cannot, however, be mastered by car designers alone – the state and industry as a whole must also contribute. The reduction of engine weight through using aluminium should also be seen in connection with the increased cost and the increased weight resulting from the extra damping which these lighter engines require.

Another similar problem which needs to be examined carefully is the use of thinner windows. On the Golf, 5.6 kg could be saved in this way if the windows at the side and rear were to be reduced from 4.2 mm to 3 mm thickness. Against this saving which, incidentally, involves extra cost, one must put extra weight for damping materials. To what extent this measure will change heat transmission will have to be ascertained by tests.

As far as the use of plastics windows is concerned, there are still many obstacles. The biggest problem is probably that the change-over from glass

Fig. 5. GRP sound insulating shield for the Diesel motor of the VW Passat.

to plastic will require very many official tests and comparisons before it is sanctioned. One should, however, remember that a change-over only of the triangular window at the front and the side window at the rear would already save 3 kg in the two-door Golf for a window thickness of 4.8 mm. Heat transmission would thereby be reduced and better soundproofing be achieved.

That weight savings do not always mean extra costs if the properties of the material are made full use of, is shown by the development of the sound-proofing shield for the Diesel motor of the VW Passat, shown in fig. 5. For technical reasons the shield was first made by welding together three deep-drawn steelplate parts, painted and given a layer of foam to reduce noise, fig. 6. By changing over to a one-part black GRP shield made of SMC (sheet moulding compound), the need to apply an anti-rust paint is eliminated since corrosion resistance is ensured even if the shield gets damaged. Furthermore, it proved possible to reduce damping measures and to reduce the weight from 3.3 kg for steel to 2.1 kg.

Fig. 6. Sound insulating shield, originally used in the Passat Diesel and welded together from three deep drawn steelplate components.

A comparison of the total energy consumption for making this shield and its use in the vehicle during its lifetime shows that there is a saving in energy of about 33 % for the GRP model compared with the one made of steel, fig. 7. This shield was the first GRP component of that size to be used by VW in 1979 with about 300 units per day, fitted to the Passat Diesel. Mass production of SMC components had already been tried out in the early seventies on the hardtop of the VW Porsche 914.

material used	weight [kg]	energy necessary for its manufacture [10^3 kJ]	energy for operation [10^3 kJ]	total energy [10^3 kJ]	savings compared with steel ΔGE [%]
steel	3,3	214	862	1076	–
GF-UP	2,1	190	528	718	33

1kJ = 0,000 278 kWh = 0,278 [10^{-3} kWh]

Fig. 7. Total energy consumption for making a motor shield and for operating the vehicle during its life.

In developing the plastics fuel tank for the Passat Variant (1973 model) the first thought was to technically improve the model then current. The plastics fuel tank, made from high density polyethylene (HDPE) became a reality and proved to be economic in manufacture, no extra costs being involved. All this was achieved not only through intensive development work but also through well-prepared production techniques. Since then, more than a million such fuel tanks, with a net weight of 3.7 kg each, have been blow moulded, see fig. 8. The plastics fuel tank is thus 1.4 kg lighter and holds nearly 5 litres more than a comparable metal tank. Today, after nearly 7 years, the advantages of plastics fuel tanks over those made of metal, realised as important at the time, are still valid. These are greater design latitude, greater capacity, lighter weight, no risk of corrosion, better heat insulation and better performance in safety tests. But it is not only these advantages which will make plastics fuel tanks increasingly important – it is their long-term serviceability. There is no danger of rust working its way underneath a paint film if the bottom of the tank gets accidentally damaged in rough terrain.

Air resistance factor and fuel consumption

Savings in fuel consumption may be effected through structural changes such as additional front and rear spoilers, which reduce the air resistance factor, the so-called c_W value. This expresses the aerodynamic quality of

Fig. 8. Blow moulding plant with a plastics fuel tank.

a car. A vehicle with a lower c_W value will also have less air resistance, fig. 9.

The rear spoiler an the Passat KLI, fig. 10, for example improves the air resistance factor of the limousine by 2.4 % from 0.420 to 0.410. The decrease of c_W by 0.01 results in a drop in fuel consumption of 0.07 litres per 100 km (average figure for fuel consumption at speeds of 90 and 120 km/h and a town cycle).

Recent tests have shown that a reduction in weight for this type of vehicle of 100 kg reduces fuel consumption by 0.3 litres per 100 km, fig. 11.

With the present rear spoiler design – integral PU foam with sheet metal insert – one must, however, put up with a weight increase of 4 kg. If one

W = air resistance
c_W = air resistance coefficient
A = projected vehicle area
ϱ = density of air
V_F = speed
P_W = air resistance capacity

$$W = c_W \cdot A \cdot \frac{\varrho}{2} V_F^2$$

$$c_W = \frac{W}{A \cdot \frac{\varrho}{2} V_F^2}$$

$$P_W = W \cdot V_F = c_W \cdot A \cdot \frac{\varrho}{2} V_F^3$$

Fig. 9. Schematic illustration of the definition of the air resistance coefficient, the c_W value for the Golf.

now puts this weight increase side by side with the fuel saving, the vehicle would have to be reduced by about 24 kg. By using a rear spoiler therefore, a theoretical weight reduction of 20 kg has been achieved.

Now it was up to the engineers to re-design this rear spoiler and to make it still lighter trough using other materials. By using high density polyethylene

Fig. 10. PU rear spoiler with metal insert on the VW Passat GLI.

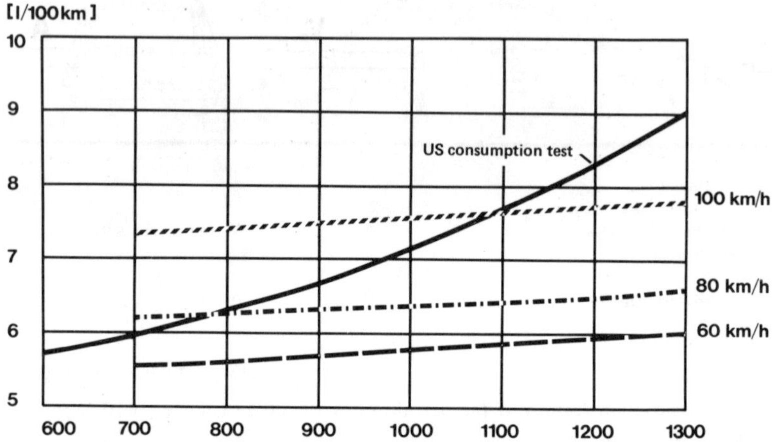

Fig. 11. Effect of vehicle weight on fuel consumption.

(HDPE), it was possible to reduce the weight of the spoiler to 500 g and at the same time achieve improved resistance to head-on collision. Tests have shown that, compared with other materials, blow moulded thermoplastic components are far better in this respect, see fig. 12.

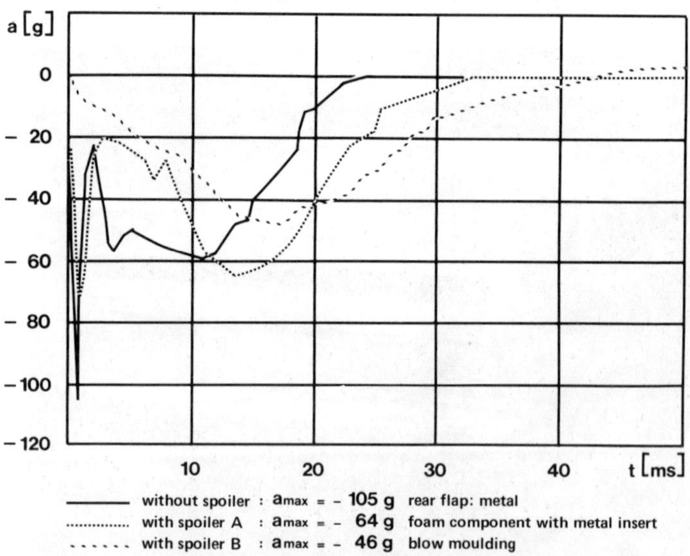

Fig. 12. Comparison of energy absorption in a head-on collision with the rear of the vehicle, for rear spoilers made from verious materials.

The maximum retardation for a metal flap without spoiler is 105 g and can thus cause serious or even fatal injuries. A mounted PU rear spoiler with metal insert an the other hand has a maximum retardation of 64 g. The lowest retardation is achieved with a thermoplastic rear spoiler, namely 46 g.

In future it will be the task of vehicle designers to revise existing ideas and designs in order to achieve lighter components with the same or even better material properties by carefully designed measures as, in the present instance, the use of other types of plastics.

What about carbodies?

As far back as the spring of 1978, special mention was made at an international conference on the use of plastics in motor vehicles, of the significance of using GRP materials for external car components. This conference laid down the criteria for continued development and gave a stimulus for future activities to follow certain directions. It is evident that much progress has been achieved since then, especially in painting techniques and formulations for SMC's with improved impact resistance. Some spectacular results have recently been published concerning prototypes such as the world record test vehicle by Daimler-Benz with a plastics body made from carbon fibre and glass fibre reinforced polyester resin, fig. 13, and the VW research

Fig. 13. Mercedes world record test vehicle C 111/III with a carbon and glass fibre reinforced polyester body (c_W = 0.195).

Fig. 14. VW research car.

car, fig. 14. The underlying thought as regards this compact VW car was to use corrosion resistant materials at those spots of the body most at risk from corrosion, and thereby make an effective contribution towards corrosion protection and reduce the effort and expense of carrying out additional anti-corrosion measures, see fig. 15. It does, however, seem important that

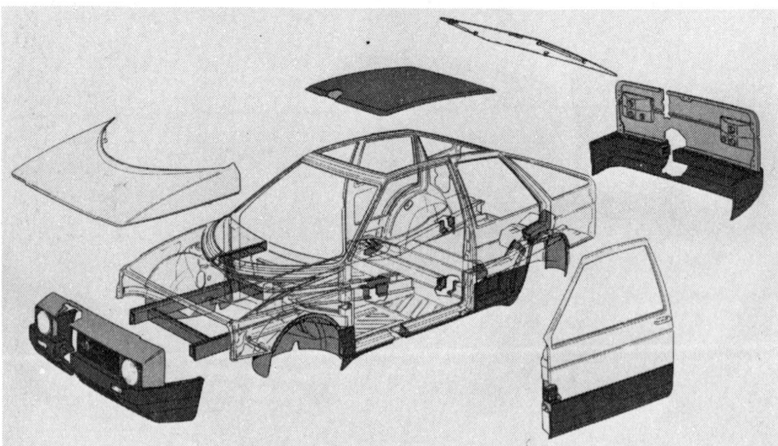

Fig. 15. Body of the VW research car with possible component groups made from plastics materials.

such ideas are put into practice on a large scale, e. g. as in Ford's small van, the Econoline-VAN with its plastics bonnet of which more than 100000 units are made annually.

World record vehicle, research car. bus developments of a technical institute, carbon fibre reinforcement, unit construction – all these are very likely only the beginnings of the path towards a new development phase, although certain difficulties still exist, such as methods of calculation for the solution of these problems. There have been frequent indications recently that it is possible to help in the change-over from sheet metal to plastics components through mathematical calculation or computer aided design (CAD). The results obtained so for are not very convincing however. We, too, have found that it is absolutely essential to support the knowledge and experience acquired over a period of so many years by mathematical techniques and to apply this in the development of plastics components in collaboration with plastics experts.

If one considers a rear cover made experimentally from GRP, strengthened with struts (calculated), real doubts occur as to whether this is the right method, see fig. 16. No designer with years of carbody design experience would ever place struts on his metal components in such a way.

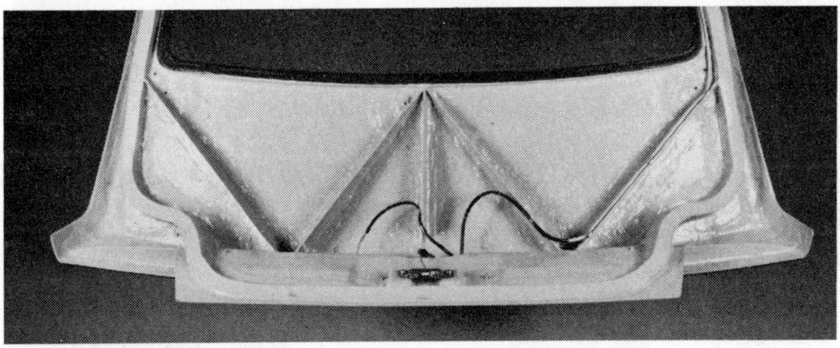

Fig. 16. Sample of a GRP rear cover showing reinforcing struts.

The positioning of reinforcing struts has always been a difficult problem and has been closely examined in technical institutes. Fig. 17 shows a suggested system of reinforcing struts for a rectangular panel, which is already being used to strengthen petrol cans. One also misses that the possibility of equal-strenght supports, which can be achieved far more easily with plastics than with steel parts, One should also study the deep drawing performance of

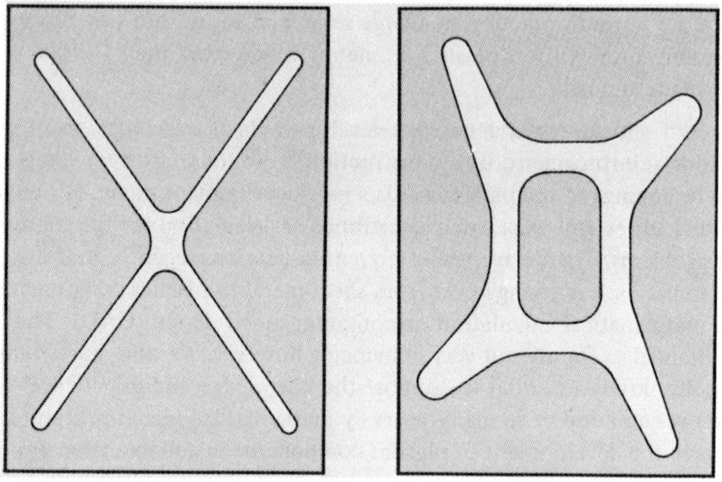

Fig. 17. Reinforcement of rectangular panels.

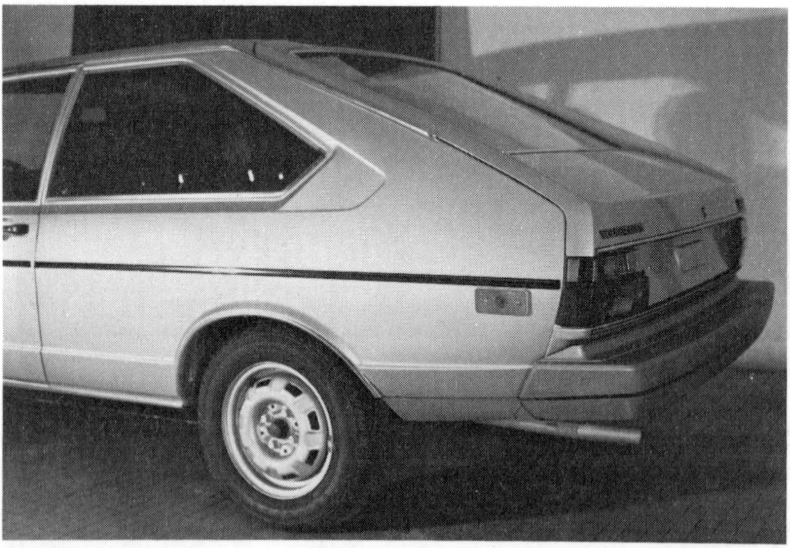

Fig. 18. Rear side window with additional reinforcing strip an the VW Passat's slanted rear.

metal sheets in tools in order to be able to understand better the behaviour of SMC during fabrication. Here, many creases which could cause problems, and compressed areas which can contribute towards reducing overall strenght, could be avoided.

It follows, therefore that it is not only development engineers who must pay attention to these problems, but production engineers, too. This collaboration already exists in the sheet metal sector. The rear side section of the Passat's sloping back, which has now got an extra strut between rear and side window to enable the metal to be more easily drawn and prevent creasing, is a good example, see fig. 18.

Although it is often explained what demands a plastics component has to comply with, it has not yet proved possible to get stylists' sympathy and understanding for the situation, despite the fact that – as has been shown in the Passat – correct design, related to the material, results in simpler manufacturing technique. This will be one of the main tasks in the near future when thought is given to use SMC parts made from polyester resin at least on component units.

Fig. 19. Plastics mudguard (PU made by RIM) used for test purposes on the VW Beetle Cabrio.

Must CMC's be made from UP resin?

Recent developments in the polyurethane sector, seen particularly clearly at the plastics exhibition K'79 in Düsseldorf, now have very bright prospects for their more widespread use, because of the excellent properties of the material. Whether PU is suitable only for small runs where parts are quickly made, using simple tools, or whether it is also suitable for large scale production, will be evident after further tests have been carried out. VW has been testing a PU mudguard, made by RIM, on its Beetle and results so far are promising, fig. 19.

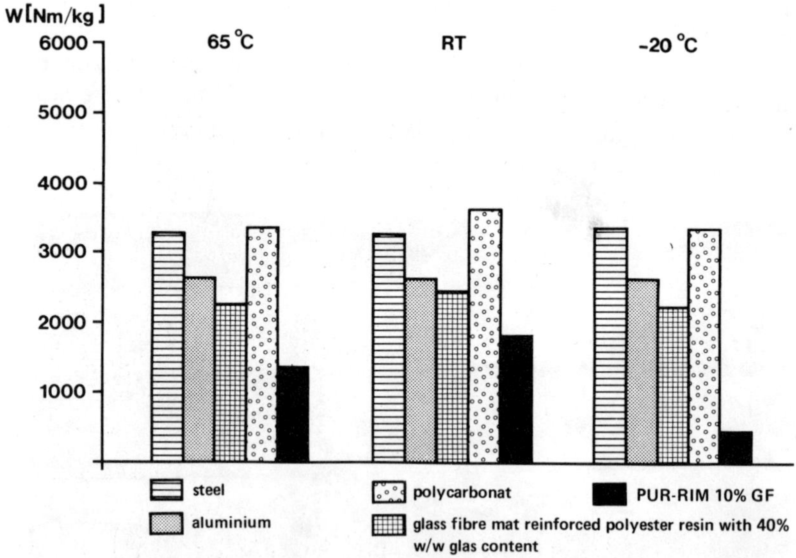

Fig. 20. Comparison of energy absorption of verious materials by the ball identation test at a constant deformation stroke of 165 mm.

Ball indentation tests have been used to find suitable alternative materials for external carbody components. These impact tests examined the splintering behaviour and energy absorption at a constant deformation stroke at temperatures of – 20 °C, room temperature RT = 20 ± 2 °C and 65 °C. The diagram shows the figures which permit comparable designs for all the materials examined, fig. 20.

External shape, aerodynamics and improvement of c_W value

Plastics front and rear spoilers, as well as underbody sealants, are useful in improving the c_W value. Much more important, however, would appear to be the possible help in the realization of styling details involving much deformation. Here we are thinking mainly of the front parts which, being made in one piece, anticipate many detail improvements in joints and connections and, moreover, guarantee perfect shapes according to stylists' requirements, in collaboration with aerodynamic experts working in wind tunnels. The medium range sports car BMW M1, fig. 21, for example,

Fig. 21. Medium range sports car BMW M1, body made from a tubular frame with a GRP outer skin.

Fig. 22. Porsche 928 with flexible front and rear cladding made from reaction mouldes PU.

whose body consists of a tubular frame with an outer GRP skin, as well as the Porsche 928, fig. 22, which has a flexible front and rear cladding made from reaction moulded PU were developed in this way. To make these from metal would very likely require the assembly of many different parts which, moreover, have to be cleaned and polished. The new models of Daimler-Benz's S-class demonstrate that if the shape has been carefully designed from the aerodynamic standpoint, a good c_W value (0.36) will be achieved.

Plastics are also making a major contribution towards solving questions of detail, such as the covers for rain channels, the improved design of rearview mirrors, fig. 23, and in bonding windows which can thus be better adapted to the outer contours of the vehicle so as to reduce air resistance.

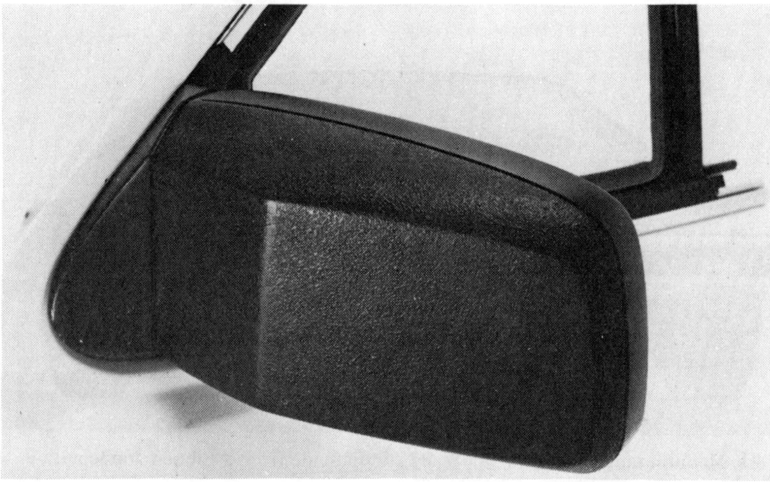

Fig. 23. Outside rearview mirror for the VW Passat, well designed from the aerodynamic point of view.

Why are plastics given a chance in production techniques?

Increased importance has lately been attached to improvements in production techniques. These relate to humanisation at the place of work, the use of structural units to cope with the multitude of different models, as well as to the trouble-free production of vehicles in different places.

Humanisation at the place of work – i. e. making life more pleasant for car assembly workers – has involved a great deal of time and effort on the part of development and production staff. The many new demands that are made on a vehicle often result in restrictions at the place of work. These restrictions should be eased by dispersing structural units which, however, would require more space in the assembly sheds. In this way one would also prevent overhead and heavy assembly work. To install lightweight and easier to assemble components is not only a question of energy for operating the vehicle. The next few years will show considerable progress in this direction.

More comfortably equipped cars also increase the number of structural units to be assembled. This is well illustrated by the increase in electrical equipment. The 1960 model of the Beetle required 60 m of cable, whereas the 1979 Passat requires 200 m. This does not yet take into consideration the fact that even more cable will be used in the near future for electronic units for assisting steering and braking, central door lock, electrically operated sun roof, electric window openers etc.

The above remarks show that plastics must be given preference and that certain problems can be solved only through using plastics.

Product systems result in greater production flexibility if one considers structural units. In this way, too, supply flexibility and production would be simplified in the various production plants. Furthermore, the effect on after-sales service and the stocking of spare parts is important.

A variety of models from one basic body

The Golf – an international car

The problem of designing an international car became very evident when the Golf was introduced in the USA as „rabbit". All designs and documents for this car were supplied by the car research department at VW's Wolfsburg works. Research personnel from the parent company were transferred to VWoA and their task was to adapt the vehicle to American production methods, taking into account

American suppliers

the wishes of American customers.

Later on, the Golf Mexico „Caribe", whose component parts originally came from Europe, was included in the American production programme. This is how the international car, the Golf, was developed, based on a basic frame, fig. 24. This basic body, which is exactly the same for every car of this type troughout the world, can be used for all models, i. e. the Golf

Fig. 24. Basic body for the VW Golf.

(Germany), Rabbit (USA), Caribe (Mexico), Golf (South Africa), Jetta, Scirocco, Golf Cabrio and, for the front part also the Rabbit Pick-up (USA), fig. 25.

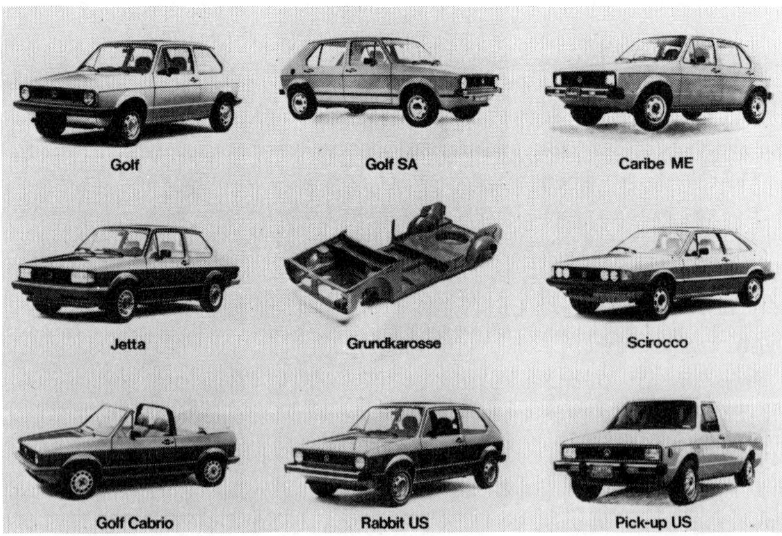

Fig. 25. Many different models of the international Golf, based on one basic body.

How are the others managing?

Let us have a look at a recent model by General Motors, the X-Car. The development of four ranges, based on the X-Car basic model, fig. 26, offers 16 different, compact types of vehicle, whether in the form of the Chevrolet Citation, the Oldsmobile Omega or with the characteristic sheet metal body of the Buick or Pontiac, fig. 27.

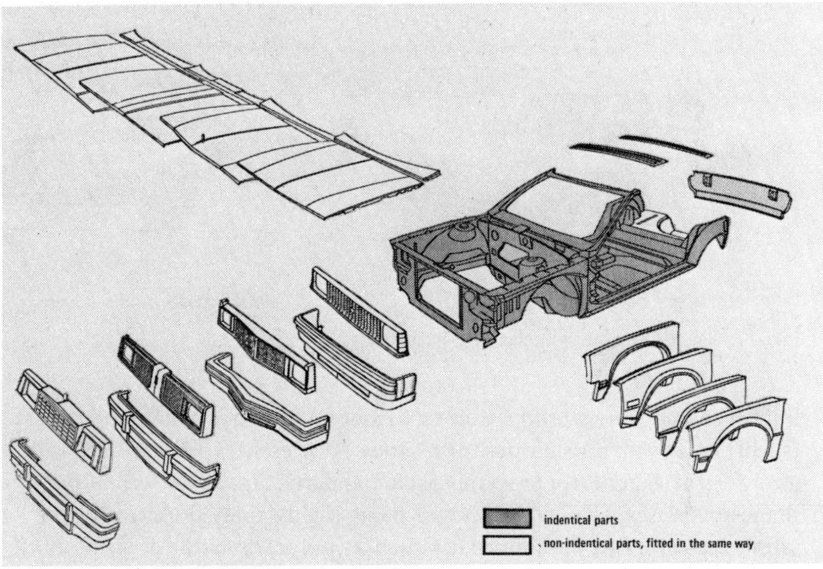

Fig. 26. GM X-Car base body with parts which can be built on.

This shows that with four different tops, mudguards and front ends, numerous vehicle variants can be produced. It is precisely these parts which can be of special significance as far as the use of plastics is concerned. This differentiation also means that the number of units in the entire group is divided into individual groups. This would offer new opportunities for plastics components which so far are only being economically produced at the rate of 200000 p. a. compared with steel parts. Moreover, one can assume that a facelift could be effected on such vehicles at less cost, since the basic body is not affected, so that metal assemblies or separate components need not also be altered. Front and rear ends are particularly suitable for the use of plastics.

Fig. 27. GM X-Car basic vehicles on a basic body.

Improved c_W values, production techniques and easier handling during a facelift offer new possibilities for plastics. At present, GRP parts are widely used as replacements for pressure-cast metal parts, fig. 28, as well as polyurethane soft-noses, fig. 29, which are being highly praised especially for safety reasons, and thermoplastics such as polycarbonate for front ends, fig. 30. Experts indicate that, after comparing the various materials discussed, preference should be given to the PU soft-nose. There are no signs as yet whether this applies only to American vehicles or whether the trend will

Fig. 28. SMC bumpers for the Porsche 924.

Fig. 29. Polyurethane soft-nose of the PONTIAC FIREBIRD (General Motor).

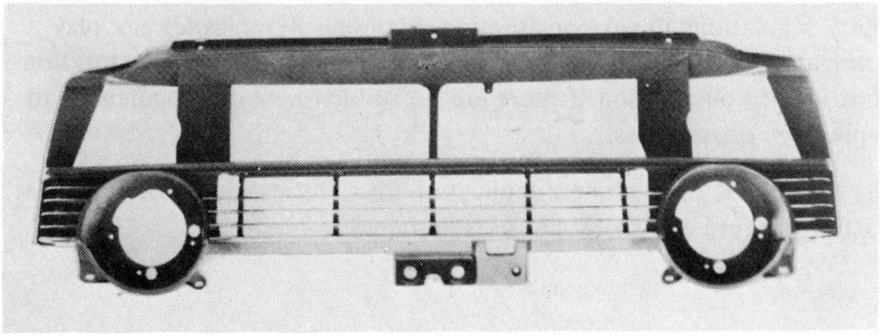

Fig. 30. Front part of the FIAT STRADA (USA) made of polycarbonate.

also continue in Europe. This will very probably depend on the design principle of the vehicle, the development objective of the manufacturer and on the serviceability of the various designs.

Summary

Plastics are highly promising materials for the car industry since they meet the new demands of vehicle construction. They are not, of course, able to replace every material, nor, indeed, are they supposed to do so. There are, however, a number of points which speak for their continued and successfull use for a wide range of components. These are:

They give stylists and engineers the possibility of producing vehicles with favourable c_W values.

They are the lightest materials for front and rear ends and additional components such as spoilers.

The use of plastics enables the most complex assemblies to be easily produced as one unit.

They protect passengers from excesses of temperature as well as reducing noise levels.

They help to dispense with anti-corrosion measures.

When industrial development produces a new material, this can replace traditional materials up to a point. This replacement progresses slowly, then more quickly, passes through a turning point and then follows a gradual, horizontal course indicating saturation of the market. This saturation limit applies only to existing developments. For additional developments, e. g. the demands made by increased passenger and driver comfort and safety regulations in car construction, materials like plastics can play a dominating part. Nevertheless, one must reckon with a certain saturation limit in car construction if there are no sudden new developments and application possibilities.

The time is ripe for such new application possibilities and it is certain that plastics will find ever increasing uses in the vehicle sector.

Bumpers and bumper covers of Daimler Benz's new S-class

Reinhold Heim

Bumpers as well as bumper covers of the new S-class were designed as one functional unit already when the first model was conceived.

The design includes, in one unit, all the parts which used to be fixed separately, e. g. impact absorbers on the bumper, bumper covering, covering under the bumper, air ventilation grille and covering of towing hook. Figs. 1 and 2 show the former design, figs. 3 and 4 the new one.

Combining these separate components into one unit necessarily led to the demand for maximum flexibility and resilience, to be achieved through using new materials, so as to reduce the risk of damage to the vehicle.

New materials allow greater scope for design, but they should also help to improve certain other features such as the c_W index and ventilation of the engine compartment, protection of the driving components, protection against falling stones and corrosion, and finally weight reduction.

Fig. 1. Front bumper and bumper covering on the old model.

Fig. 2. Rear bumper and bumper covering of the old model

Attemps were also made to prevent, as far as possible, permanent deformation of front and rear spoilers through contact with the road surface and in minor accidents. Once the styling model had been approved, a specification was drawn up, listing the required functions of the component after being fitted to the vehicle. Here, the most important considerations for bumper covers were:

Fig. 3. Front bumper and bumper covering of the new model

Fig. 4. Rear bumper and bumper covering of the new model

1. As accurate a reproduction of the model state in the vehicle.
2. Maximum flexibility and resilience at all mechanical loadings down to – 25 °C.
3. Sufficient rigidity – after all, a self-supporting steel plate car body component has to be replaced.
4. The paint used should give a film that is at least as flexible as the painted component.

Potentially useful materials were examined for suitability, bearing the above points in mind. Here one must also remember that material selection and specification had to take place about years ago so as to have plenty of time available for trials before they could finally be passed for mass production. This, of course, meant that the choice was limited to materials available at that time, i. e. EPDM, modified polyolefins, thermoplastic polyurethane elastomers, thermoplastic injection moulding compounds and structural polyurethane foam.

First of all, a well-known and proven material was examined, namely EPDM, a synthetic, cross-linked rubber elastomer which has already been used for the impact absorbers on bumpers. This material had, however, to be excluded for the intended application because it is relatively heavy and not stiff enough. This meant that, in order to achieve sufficient rigidity, the section thickness of the component would have had to be increased, which would have made it too heavy. Furthermore, there was but insufficient experience as regards the painting of EPDM components and no lang-term test results were available.

The second group of potentially useful materials was elastomer modified polyolefine, which are already being used by a number of car manufacturers, in most cases as a covering for bumper brackets, as impact absorbers on bumpers or for bumper corners. These are very good examples of the use of these materials because the components are not self-supporting, have not, so far, had to be painted and can be made with a textured finish to hide surface defects.

The disadvantages of these materials and thus the reasons why they are not employed, are inadequate heat distortion resistance for a self-supporting car body component, smooth, unpainted components are susceptible to scratching and a satisfactory paint has not yet been discovered. Recent developments both as regards material characteristics and paintability do, however, indicate further applications for these materials.

The third group of materials to be examined was that of the thermoplastic polyurethanes. The low temperature impact resistance of the polyurethanes on the market at that time was still inadequate and it was not until 1978 that grades were introduces whose low temperature impact strength and other mecanical properties were comparable to those of structural PUR foam. Before these new grades are used for mass production however, extensive tests will have to be carried out to ascertain their long-term performance and paintability. It is probable that the raw material cost, which so far is still grater than for PUR structural foam, will be partly compensated by technical advantages. Advantages, for example, include the fact that no mould release agent is necessary and that scrap can be re-used, better cycle times depending on the type of moulding being made and, possibly, paintability without prior application of a primer.

Whether major applications for PUR made be expected in the future depends upon the results of the above mentioned tests, on questions of profitability and last but not least on the continuing development of these materials, including the possibility of painting them.

The fourth group of materials to be examined for this purpose was that of the "classic" thermoplastics such as ABS, PP, nylon and thermoplastic polyesters. They, too, had to be rejected, partly on the grounds of indadequate heat distortion resistance, partly because of insufficient low temperature impact strength but, above all, because the shatter unter high speed impact.

Recent developments, e. g. nylons and polycarbonates, which are claimed to effect a distinct improvement of the above mentioned disadvantages, will have to be examined and could, perhaps, open up further application possibilities in the car body field.

After the above mentioned materials had been examined there now only remained structural polyurethane foam, a material which was, of course, known to us already from its use for the bumpers of our USA models and from our current 200 – 280 models as bumper corners and headlamps. Here, the demand as regards stiffness and paintability were not so exacting for the intended application.

Existing PUR formulations therefore had to be improved as regards stiffness, heat distortion characteristics and paintability. In the time still available this was possible only because heatable epoxy resin moulds are adequate for making quality components. Such moulds are made relatively quickly and cheaply – an advantage offered by no other group of materials. The other reason was that all those taking part in the project, e. g. raw material producers, fabricators, the paint industry and our own staff, worked as a team.

In this way, new paths were trodden as regards the design of the front and rear cladding of our new S-class. The new designs brought weight savings compared with the previous model and a distinct improvement in performance. The question nevertheless remains open whether, bearing in mind the many new aspects and requirements, such problem solutions will ever be repeated in the future. It is certain that this PUR system is unsurpassable with respect to performance, design and surface finish, as well as the possibility of matching properties to a particular application. Future applications will not least also be a question of profitability and, compared with other materials, future prospects for polyurethanes may be regarded with hope.

Bumpers of Daimler-Benz's new S-class

Wolfgang Klie and Wolfgang Fischer

General demands made on a bumper system – boundary conditions

The main function of bumpers is to protect the front and rear of a vehicle against damage. There are considerable differences of opinion as to how effective this protection should be, i. e. how much damage is permissible at what speed of impact.

Since, for obvious reasons, there is no statistical information about the number of accidents in relation to speed of impact (and even less so for slow speeds) it is difficult to carry out economically relevant cost/profit analyses on this question. For this reason, the purpose behind the development of our new bumper system was first formulated in a general sense, as detailed below.

a) The requirement stated in ISO 2958 had to be reliably fulfilled.

b) The weitght had to be reduced compared with the previous model, and its function improved.

c) Tests had to be carried out to find out whether the European bumper system could be standardised with the USA system which likewise has to be newly developed.

d) The design requirement and, particularly, the aerodynamic requirements as regards improvement of the c_W-value and the control of the cooling air had to be taken into account.

System layout Europe – USA – energy considerations

A bumper system normally includes a bumper bracket and one or more energy absorbers. Two designs are possible: the bumper bracket is either in front of the absorber system or behind it, fig. 1.

The aimed-at weight, the possible projection of the bumpers beyond the body (vehicle length) and the load to which a car body can be subjected at the front and rear ends – these are the important considerations. The possible impact speeds for a given vehicle mass are governed by the

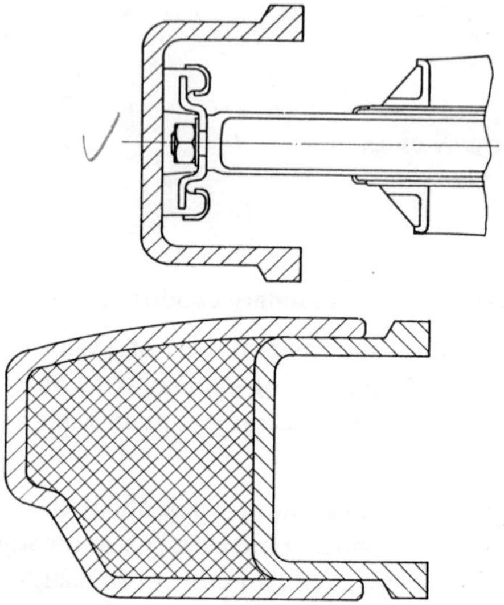

Fig. 1. Bumper construction

variations in the quality of the energy absorbers and the mechanical properties of the materials that could be used for the bumper brackets. Diagr. 1a – d illustrate the relationships and one or two still open questions can now be easily answered:

Standardisation of European and USA bumpers?

Increasing the possible impact speed from 3 km/h to 8 km/h on to a rigid object is achieved only by making the bumper bracket alone more than 10 kg heavier (diagr. 1d) and increasing the vehicle length by more than 160 mm. To this must be added the considerable extra cost for the hydraulic shock absorbers. Standardisation is therefore not possible, especially in view of the absolute priority of weight reduction.

What kinds of absorber systems should be employed?

For the USA design, only hydraulic shock absorbers with a quality factor of $\eta \geqslant 0.85$ can be used. So far, there is no plastics material which will reach this figure.

For the European system it is necessary to develop plastically deformable steel strip supports which should have a quality factor of $\eta \geqslant 0.7$.

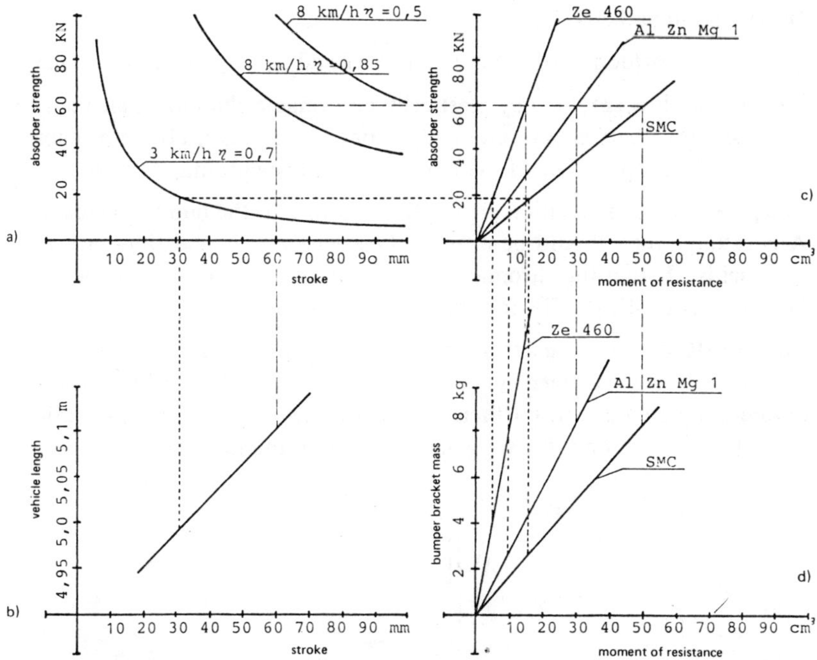

Diagram 1a – d. Relationship between vehicle length, absorber strength and bumper bracket mass.

a absorber strength as a function of stroke
b vehicle length as a function of stroke
c absorber strength as a function of moment of resistance
d bumper bracket mass as a function of moment of resistance

What materials could be used for the

Micro-alloyed steel plate, aluminium alloy extrusions and glass fibre reinforced plastics could be used. Using quasi-static flexural tests on U-profiles, we were able to determine the flexural strengths of the different materials for our investigations. Diagr. 1c and 1d show that aluminium and SMC profiles will be about the same weight, steel plate profiles on the other hand heavier.

Development variants

The basic construction of the bumper was fixed already at an early stage.

The bumper bracket, held by plastically deformable aluminium elements or by shock absorbers is covered on the front by an elastically deformable cladding. The top surface of the bumper should be chromium plated.

The possibilities of variation are confined mainly to the bumper bracket. A question which should be examined especially in connection with the material is whether the chromium plated top should be designed as a load-bearing part of the profile or purely as a means of masking the bracket.

For an SMC bumper bracket it is necessary to provide a separate chromium plated cover whereas steel and aluminium profiles can be partly or wholly chromium plated direct. Furthermore, a two-shell method of construction is possible, in which only the top part need be chromium plated.

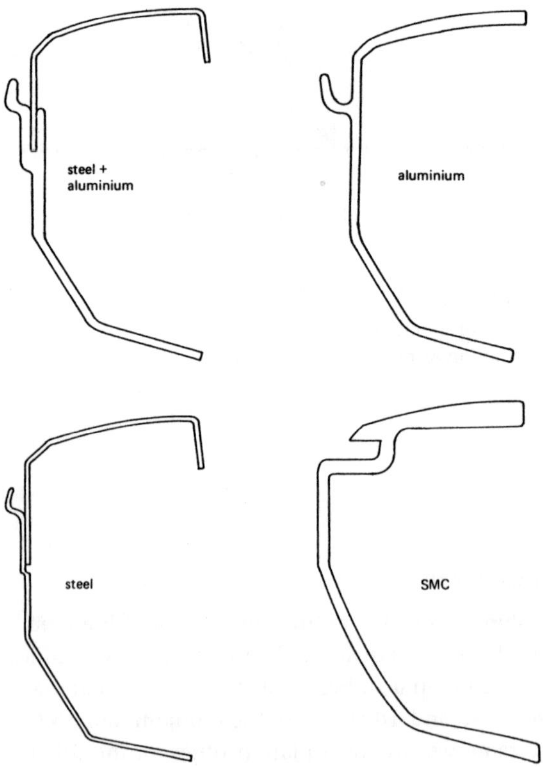

Fig. 2. Profile comparisons for European design

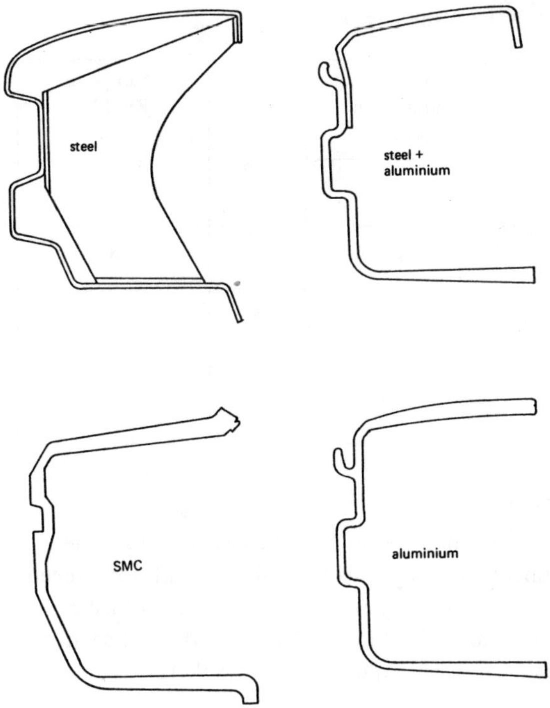

Fig. 3. Profile comparisons for USA design

A combination of two materials has also been discussed, variants in figs. 2 and 3.

Assessment of variants
Reasons for selection

Table 1 lists the various materials which are suitable and assesses them according to their most important characteristics:

Flexural strength

If one considers the flexural strength in relation to cross-section area (diagr. 1c and 1d), it is evident that the extruded aluminium profile is about the same as the SMC profile. The steel profile, on the other hand, is clearly inferior.

Table 1. Assessment of material properties for

Property	Extruded aluminium profile	SMC	Steel plate ZE 420
Flexural strength/ cross-sectional area	+	+	–
Introduction of forces/ dents	+	–	0
Flexibility	+	±	+
Breaking strength	+	–	+
Corrosion, surface	–	+	–
Component design	–	+	–

Introduction of forces; dents

According to experience so far, there are no problems as far as extruded aluminium profiles are concerned. Since steel profiles should not exceed 2 mm material thickness because of weight considerations, denting can only prevented by means of reinforcing ribs. Fig. 2 shows the effect of section thickness on the dent resistance of U-shaped bumper brackets.

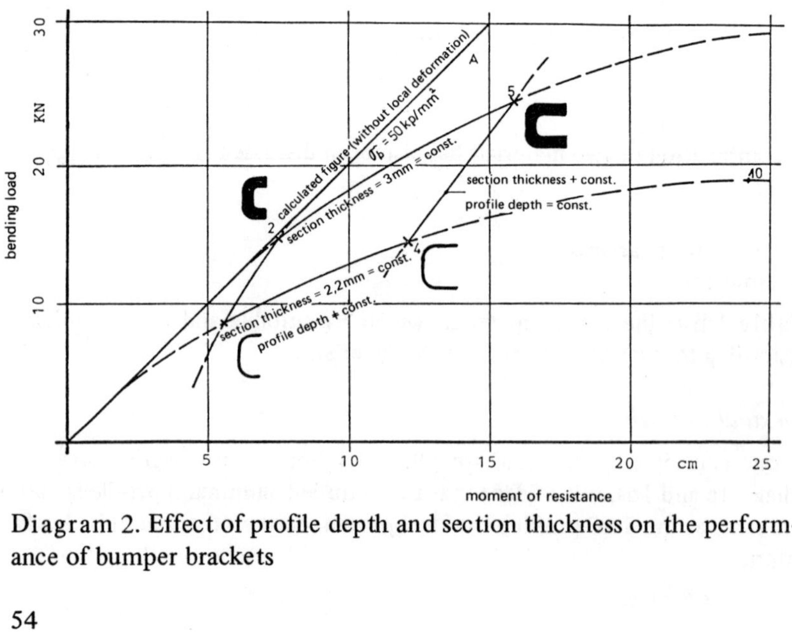

Diagram 2. Effect of profile depth and section thickness on the performance of bumper brackets

In the case of SMC, additional measures have to be taken since good flexural strength can otherwise be achieved only with difficulty.

Flexibility and breaking strength

The great flexibility of SMC is advantageous if there is sufficient space for the large deflection of the bracket. It can then be used direct as a spring element and absorb shocks. If there is insufficient space, the SMC bracket must be dimensioned for deflection, which could possibly involve extra weight.

Since SMC allows practically no plastic deformations, it is likely that the bumper bracket will break if the maximum bending load is exceeded even slightly, whereas in the case of aluminium profiles the plastic deformation is even less.

Corrosion/surface

The position of bumpers on the vehicle means that the material of which they are made must be extremely corrosion resistant. Compared with steel and aluminium, SMC meets this requirement most readily. Since the profile has to be completely covered in any case, there are no problems regarding the finish of the painted surface.

Component design

The thickness of steel plate is always constant across the entire width. In the case of extruded aluminium profiles it can vary across the cross-section, whilst in the case of SMC, it can vary transversely as well as longitudinally. This means that the section thickness of SMC components can be varied as required by the critical loads likely to be encountered.

Results

The expected cost advantages for steel were no so great, especially in view of the additional measures (struts), in relation to the extra weight that had to be accepted. Steel was thus excluded at an early stage.

Combinations of different materials, e. g. an aluminium top part and an SMC bottom part were not investigated further because it seemed likely that the considerable differences in elastic modulus would cause problems.

The expected cost differences between SMC and aluminium were not so marked that an unequivocal decision in favour of one or the other material could be made. Furthermore, advantages and disadvantages more or less cancelled each other out.

Figs. 4 – 7 show the solutions for the four bumper designs. For the European bumper, aluminium was chosen for the front bumper, SMC for the rear one. For the USA versions the opposite arrangement was decided upon.

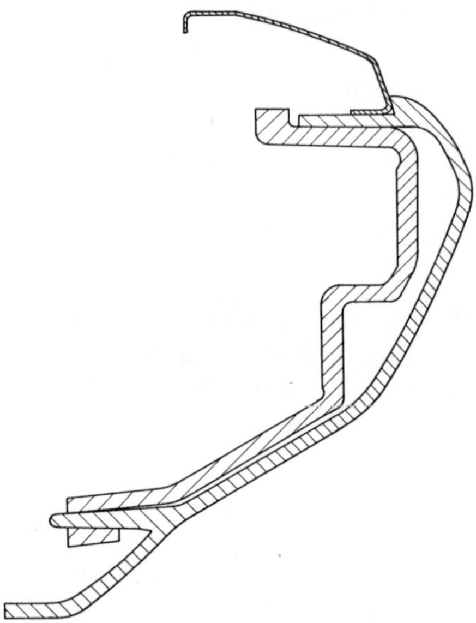

Fig. 4. Rear bumper (European)

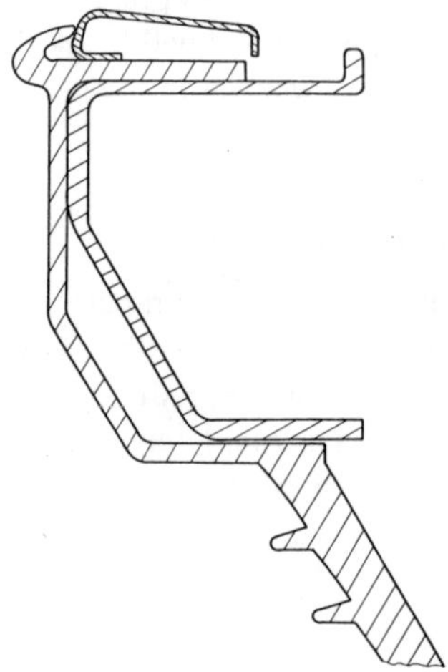

Fig. 5. Front bumper (European)

The reason for the decision is that it is easiest to acquire experience with the new SMC technology, also as regards cost trends, if SMC and aluminium can be compared direct under mass production conditions. A reversal of materials used in the European bumpers would also have been possible.

As far as the rear USA bumper is concerned, there are space problems so that SMC cannot here be used because the deflection would be too great.

There are technical reasons (headlamp washing equipment) why, despite the aluminium profile at the front a separate chromium plated covering had to be provided for the European bumper.

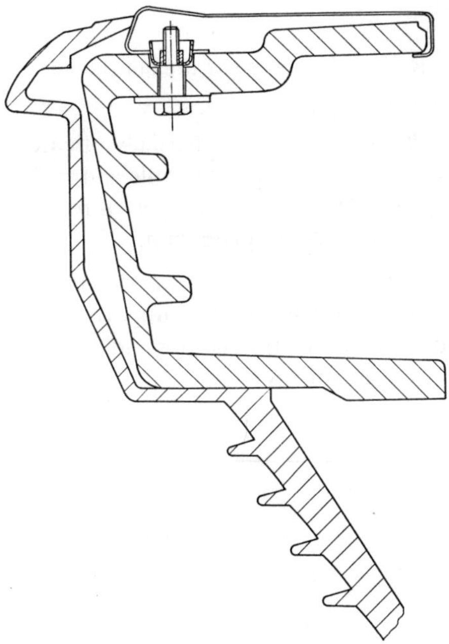
Fig. 6. Front bumper (USA)

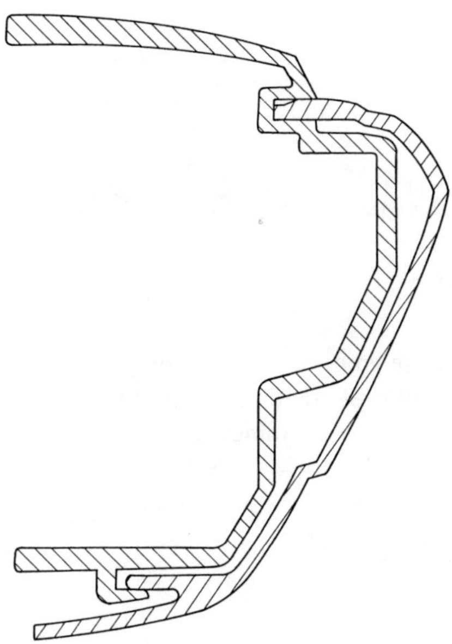
Fig. 7. Rear bumper (USA)

Series development

The following problems had to be solved in the course of the continued development of SMC profiles for the USA bumper:

The required flexural strength was easily achieved by the partially oriented glass fibres. Major problems arose at the front face of the profile and at the support points, due to problem associated with the introduction of forces. Delamination was caused especially by the USA test pendulum.

Geometric modifications of the cross-sectional pattern at the top, the section thickness at the front (horizontal ribs) and near the supports were found to be necessary in addition to optimissing formulations and assembly technique, in order to meet the requirements of US tests.

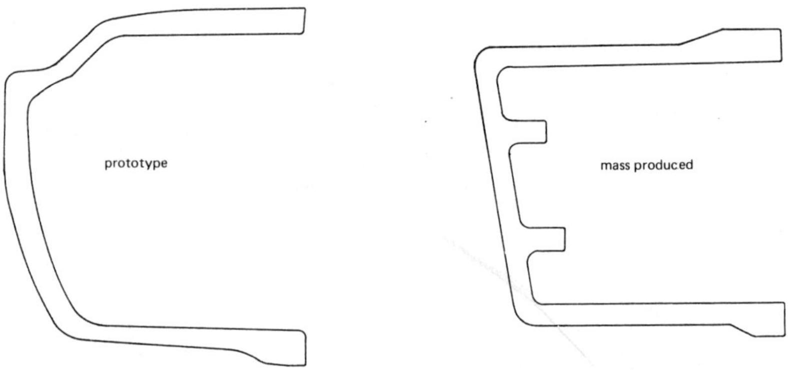

·Fig. 8. Improvement of SMC bumper (USA)

Fig. 8 shows a comparison between the mass produçed profile and the prototype profile.

Future outlook

Whether or not fibre reinforced plastics are going to be increasingly used for bumpers in the future depends largely on cost trends compared with aluminium. If fabourable cost trends were to be joined by improved material properties which would enable the weight to be further reduced and the component's function to be improved, there is no reason why plastics should not be used exclusively for bumpers.

Development of front and rear ends of the Fiat Ritmo

Claudio Noris

The Fiat Ritmo is fitted with plastic front and rear ends which, apart from distinguishing the car, fig. 1, ensure that it conforms to I.S.O. Standard 2958.

These notes aim to retrace the progress of the ends from development to mass production.

Fig. 1. Ends of Fiat Ritmo.

Development

The Fiat project specified the following requirements for the ends:

a) No damage or permanent deformation to occur using an I.S.O. pendulum [1] at 4 km an hour. Deflection on impact to be kept within 70 mm to prevent damage to power plant or body parts.

b) Front end to be sufficiently rigid to take the weight of the headlights. This was to allow preparation of sub-assemblies away from the car assembly line.

c) Colouring of ends through mass pigmentation in unified grey, suitable for matching with the whole range of body colours available.

There was no one answer to the problem of material selection since plastics cover a very wide range of properties. Nevertheless, the following general conditions were to be satisfied:

High deformation work in the elastic phase (see Appendix).

Ductility throughout the range of operating temperatures.

Permanence of the mechanical properties and outer appearance after aging.

Freedom from stress-cracking duc to contact with fuel.

Reliable transformation technology and consistent with mass production requirements.

Low weight and cost.

Possibility of re-cycling the material.

Low energy content.

Materials combining to a great extent these properties seemed to be the Elastomer Modified Polyolefins (EMP).

These were chosen also in the light of the experience gained in the field on bumpers constructed in the same material which had replaced the conventional steel bumpers on the new versions of two Fiat models (128 and 126) [2]. Similar extensions were underway for Fiat itself and other manufacturers [3].

The main objective was to identify thickness and internal geometry which would meet the project specifications whilst at the same time complying with the outer shape of the ends.

The dimensioning was much more difficult than for a bumper because, instead of a simple structure such as that of a beam, this was a 2-dimension

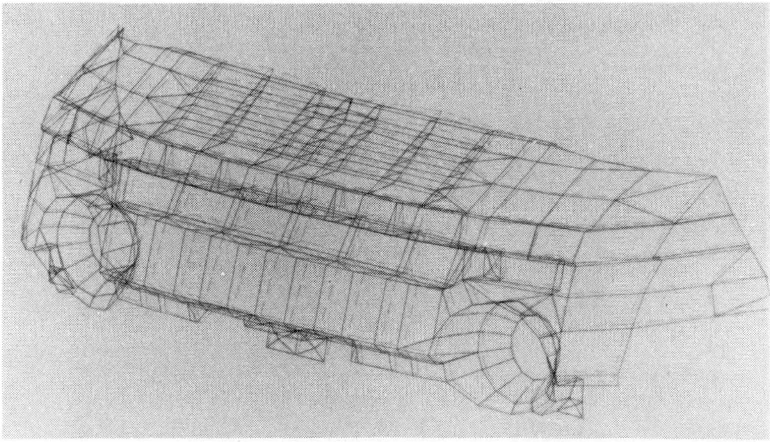

Fig. 2. Mathematical model of the front end. The broken lines represent the instant deformations on impact at 5 km/h.

structure having a complex form due to the presence of protrusions, recesses and openings. Fig. 2 shows the mathematical model of the front end [4].

The development of prototypes was thus necessary. This was made possible thanks to a pilot mould in epoxy resin suitable for injection moulding.

These first prototypes were subjected to static and impact tests which allowed the identification of critical zones containing stress concentrations. Figures 3 and 4 illustrate the test set-ups.

As a consequence, the plug of the mould was modified to bring about an increase in thickness, radius and ribbings within these critical areas.

These modified prototypes were then subjected to mechanical testing at the Stars Laboratory and a final on-vehicle testing was made at the Fiat Safety Centre.

To improve performance it was decided to include two rubber pads behind the two ends to reduce the local deflection and thereby protect the parking lights.

Once it was ascertained that specification "a" had been satisfied, the moulds earmarked for production were constructed. Accurate spotting was carried out on these moulds to optimize material flow conditions during injection, in order to obtain stress-free parts.

Fig. 3. Static test arrangement.

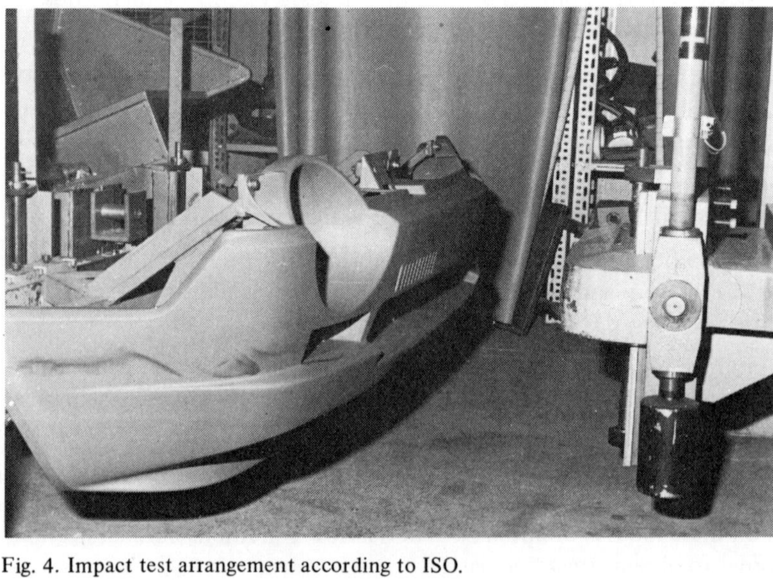

Fig. 4. Impact test arrangement according to ISO.

Production

Seen in the moulding context, front and rear end specifications are:

	front	rear
Volume in cm^3	6290	5520 (*)
Projected area in cm^2	6100	5500

(*) Weights are 5,7 and 5,0 respectively (density of material = 0,905)

On the basis of these data, injection presses were adopted having the following features:

Plasticizing capacity	15 kg/shot max
Clamping force	30000 kN max

A pressure of about 3800 N/cm^2 is applied to the material during moulding.

Productivity is 20 to 21 parts/hour.

It is obvious that the quality requirements of parts such as these call for efficient quality control during all stages of production.

This includes:

Testing of raw material

Checking consistency of machine operating conditions

Inspecting at least one specimen per production shift for the absence of damage after two successive impacts using a pendulum at 5 and 6 km/h.

A factor kept under control, being to a certain extent a measure of the consistency of operating conditions, is the weight of the part. This does not exceed ± 1 %.

Recently an important step forward took place, with the simultaneous moulding of a pair of ends.

This was achieved without using exceptionally-sized presses, not commonly available on the market.

It has been mentioned that the moulding requires a pressure on the material of about 3800 N/cm^2. Hence to mould a couple of ends in one shot, a clamping force of 44000 kN (4400 ton) is required, the total projected area being 11600 cm^2.

Using a 35000 kN machine, namely the maximum size available on the market, the problem was to fill a gap of 9000 kN. We solved this problem by fitting the moulds with additional clamping devices. These devices are operated by hydraulic jacks and are activated during the injection stage.

Compared to 1-part moulding, productivity increases from 20 to 34 parts/hour, namely an increment of 70 %.

Performance

Impact

The ends satisfy the design requirements with ample margin. By way of example, Table 1 shows the average performance of an end fitted in an on-vehicle position on a moving carriage (figure 3).

Table 1. Front End – Centred impact on lower projection centreline (a) (b) (pendulum and carriage mass = 950 kg).

Pendulum speed (km/h)	Kinetic energy (c) (Joule)	Deflection f_m (mm)	nF_m (d) (daN)
2	73	25	292
3	165	43	384
4	293	63	465
5	458	85	539
6	660	108	611

(a) One specimen per speed level.
(b) End without rubber pads.
(c) Half pendulum energy (Appendix B).
(d) Calculated using formula (1) of Appendix.

At about 7 km/h the ends crack in some areas. No shatter occurs even at the lowest operating temperature (– 20 °C). Performance at repeated impacts exhibits a fatigue limit of 5.5 km/hour for in the region of 50 impacts.

Creep

In on-vehicle position, the ends subjected for 24 hours to a temperature of 100 °C exhibit neither creep nor permanent deformation.

In particular, after the test, the headlight alignment on the front end is unchanged. This means that the original shape will remain unaltered for an unlimited period of time even in the presence of exceptional heat sources as that during the body paint touching up.

Aging

The material incorporates special stabilizing agents against downgrading due to ultra-violet radiation and heat.

The excellent permanence of the properties is proved by accelerated tests as well as direct exposure to atmospheric agents. The latter has to date amounted to more than 3 years with no measurable change in material properties.

On-vehicle performance

Fiat has carried out in-depth research on damage to the Ritmo and associated repair costs following front and rear collisions. In the first place, it emerged that no damage was suffered by the vehicle when impacted using an I.S.O. pendulum up to 6 km/hour. Furthermore, repair costs are lower than with vehicles in the same category fitted with other types of protection throughout the range of speeds examined (20 km/h). This is obviously due to the design concepts adopted for the vehicle as a whole.

Of more specific interest in our case is the clear contribution of the plastic ends in containing repair costs, if for no other reason than that they suffer no damage at impact speeds up to 8 km/hour.

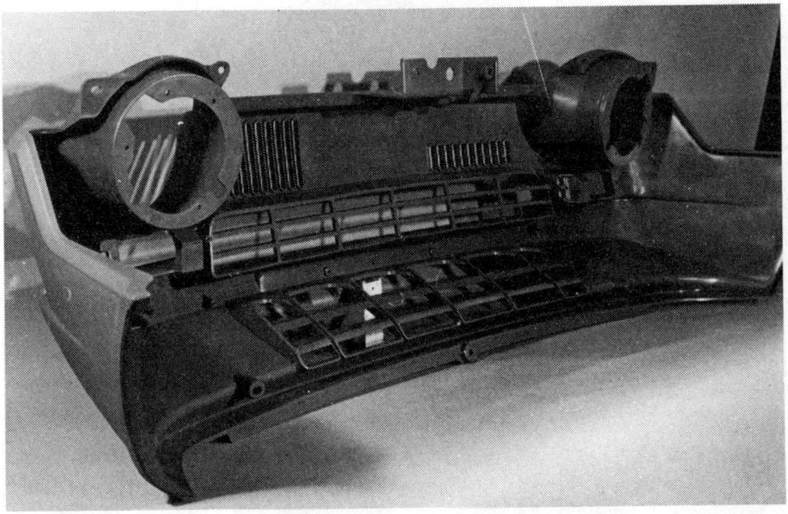

Fig. 5. Front end, inner face.

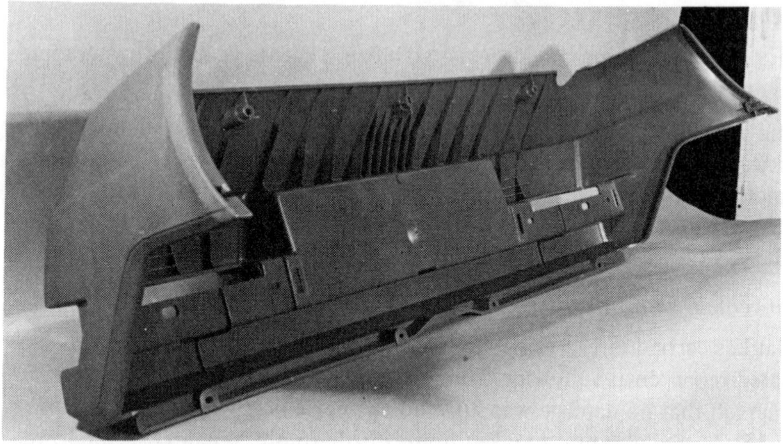

Fig. 6. Rear end, inner face.

Fig. 7. Front and rear ends coming out of the moduls.

Incidentally, it should be noted that this value is higher than that registered with the free end, due to the effect of the two rubber pads.

It should perhaps be remembered that conventional steel bumpers suffer visible damage with impacts of just 2 to 3 km/h.

Developments and forecasts

Staying in the field of the Ritmo, mention should be made of a special series called the "Targa Oro" (1979) which uses the same ends as the standard version but of a different colour (brown and black).

One application of EMP front and rear ends can be found in the Citroen Visa (1979). The front end also incorporates the radiator grille.

Current European consumption of EMP for bumpers and ends is estimated in the region of 12000 ton/year. Since 1975, when the first EMP bumpers appeared on the Fiat 128, the average growth rate has exceeded 2400 ton/year. Programmes are already being implemented by the various European manufacturers for other applications of the same type.

The colouring of the EMP ends and bumpers manufactured to date is generally standardised for the whole range of body colours and is obtained by means of mass pigmentation. This process provides for a fairly large range of light-resistant colours at low cost and with the added advantage to the user that the modest damage which may occur during the life of the vehicle is hardly visible and does not require repair.

It may be that changes in style will require bumpers and ends to match the colour of the body. In this case use will have to be made of painting especially if metallized colours are to be reproduced.

As for all the thermoplastics, paints and cycle are different than those used for sheet metal bodies.

As is known, painting of EMP's is critical from the point of view of adhesion, since highly apolar materials are involved. Nevertheless, there are paint systems available which are quite reliable and tested in the field for a number of years.

One typical system consists of degreasing with chlorinated solvents, followed by the application of a primer as adhesion promoter and a flexible enamel based on urethane elastomers.

On the other side, the manufacturers are developing modifications to EMP's aimed to supply them with built in adhesion property.

These consist in adding to the formulation special components capable of bringing to the surface their polar groups, thereby improving adhesion.

Appendix

A) The work of deformation of a structure submitted to the deflection f_m is given by:

$$W = \int_{o}^{f_m} F df$$

where F(f) is the reaction force of the structure.

This can also be expressed as:

$$W = n \cdot F_m \cdot f_m \quad (1)$$

Where $n = \overline{F}/F_m$ is the ratio between the mean value of the force and the value F_m obtained in correspondence with f_m.

Formula (1) may be given the more general form:

$$W = Hn\sigma_m\epsilon_m = Hn\sigma^2{}_m/E_s \quad (2)$$

where:

H = a factor having dimensions of a volume and depending on the geometry of the structure and on the mode of deformation.

$E_s = \sigma_m/\epsilon_m$ or secant modulus.

Within the elastic limit, the quantity $n\sigma_e/E_s$ represents the energy stored per unit volume. In the proportional field we have n = 0,5, while E_s has the meaning of Young modulus. This case applies exactly to metals.

On their part, plastics may be classified in two categories:

a) Fragile plastics in which the proportional field stops with breaking but is more extended than that of metals, reaching strains as high as 1 – 1,5 %. To this class belong the thermosetting plastics and many reinforced thermoplastics.

b) Ductile plastics in which there is no exactly proportional field where yielding is sometimes difficult to detect.
In these materials – generally low-modulus thermoplastics – the σ,ϵ curve has its concavity towards the axis $\epsilon (n > 0.5)$. Moreover, the possibility to recover the deformation, be it in the form of retarded elasticity, reaches strains of up to 10 %.

Table 2. Deformability features in the elastic field.

Material	n	σ_e (*) (N/mm^2)	E_S (N/mm^2)	$n \cdot \sigma_e^2/E_S$ (J/cm^3)
SMC with 30 % of random glass	0,5	100	10000	0,50
Polyamide 6 with 30 % of random glass	0,5	70	7000	0,35
Non-reinforced ductile thermoplastics (average)	0,6	45	2000	0,61
Elastomer-Modified-Polyolefins (EMP)	0,7	21	500	0,62
Steel sheet	0,5	210 (**)	150000 (**)	0,15
Light-alloy sheet	0,5	200 (**)	60000 (**)	0,33

(*) Maximum unit stress in the elastic field
(**) Values corresponding to a permanent deformation of 0.1 %

Table 2 shows for some selected plastics and 2 typical sheets the aptitude for storing elastic energy.

This makes clear the superiority of the former compared with the latter.

B) If the deformation is caused by the transformation of kinetic energy, this must be equal to (2). To further have the structure stay within the elastic limit, the following is necessary:

$$\frac{1}{2} mv^2 \leqslant Hn\sigma_e^2/E_S$$

i. e.:

$$v \leqslant \sqrt{2H/m} \cdot \sqrt{n/E_S} \cdot \sigma_e$$

This condition represents the impact against barrier of a car with mass m and speed v.

In the case of impact of a pendulum with mass m and speed v_p against a car of equal mass standing still and free of movement, one half only of the kinetic energy will influence the deformation, i. e.:

$$v_p \leqslant 2\sqrt{H/m} \cdot \sqrt{n/E_S} \cdot \sigma_e$$

Bibliography

[1] International Standard Organisation Std. 2958

[2] *C. Noris:* Developments in Thermoplastic Bumpers – 4th Sitev, Geneva, May 1977

[3] Elastomer – Modified – Thermoplastics. Gain in utility, Applications – Modern Plastics International March, 1978

[4] *F. Gay, A. Garro* e *P. Uslenghi:* Interventi del calcolo strutturale nella progettazione della Fiat Ritmo – Rivista ATA, May 1978

Front and rear spoilers on BMW vehicles

Rudolf Fernengel

Front and rear spoilers are used as aerodynamic aids not only in vehicles used for racing and similar competitive events, but increasingly in coupés and limousines for daily use. Correctly designed (not only according to appearance) they reduce lift at high speeds, improve the c_w value and thus enable better use to be made of the available power.

In 1972, front and rear spoilers were fitted to BMW cars for the first time as standard feature, 2000 BMW 2002 turbo cars being thus equipped, see fig. 1. For economic reasons, this relatively small number of front and rear spoilers was made by thermoforming ABS sheet.

Fig. 1. BMW 2003 turbo with front spoiler made of glass fibre reinforced low profile unsaturated polyester resin and rear spoiler made of flexible, integral PU foam.

The first driving trials with ABS front and rear spoilers very soon showed up the limitations of this material. Already after the first 10 – 20.000 km the first micro-cracks started to appear in the front spoiler, caused by constant wind pressure during driving and the associated alternating bending stresses. Once a micro-crack has formed, only a fraction of the original strength of ABS remains. This material is therefore unsuitable for the self-supporting front spoiler.

Following these negative results, a material was chosen which had already proved itself under practical conditions, namely a glass fibre reinforced polyester resin. When a start had already been made with the construction of steel moulds for moulding low-profile unsaturated polyester resins with glass reinforcement, the authorities demanded a so-called "soft" rear spoiler. The required hardness for such spoilers was initially given as 55 Shore A, measured on the rear spoiler surface.

Whilst the production of the press moulding tool for the front spoiler proceeded according to plan, work started on developing a completely new rear spoiler made from integral flexible PU foams. Here, the biggest problem proved to be the development of a suitable foam system and fixing the degree of shrinkage.

The first foam systems used had very many small bubbles on the visible surface so that, after extensive tests, synthetic resin moulds with release agent were discarded in favour of thermoformed polystyrene sheet. Flexible, integral PU foam components made in a polystyrene sheet mould not only have a clean, uniform surface, but can be painted without any surface pretreatment being necessary.

Neglecting the shrinkage produced during the production of the synthetic resin mould, we found a shrinkage of about 7 mm in the case of the thermoformed polystyrene sheet, which has to be balanced against the supporting mould with special care, and about 40 mm shrinkage on the finished component.

After the first prototype, ready for fitting, had been presented to the authorities (with a Shore A hardness of 50) they decided to change the original specification to a Shore A hardness of 35, measured on the surface of the rear spoiler. For this, a new foam system was developed, shrinkage determined once again and mass production taken up in good time. The first 500 rear spoilers were produced in BMW's plastics development department, after which an outside firm was entrusted with the task of making these components.

In developing a new spoiler generation for the BMW coupé 635 CSi, it proved possible – in the light of experience with an integral PU foam front apron, fig. 2 – to replace the GRP by a semi-rigid, integral PU foam, fig. 3, in the 7-range. Although the GRP spoiler fitted to the 2002 turbo had proved successful under actual driving conditions, it was very liable to break, especially when the car passed over kerbstones.

Fig. 2. BMW 735 with front apron made of semi-rigid, integral PU foam.

Fig. 3. BMW 635 CSi with front spoiler made of semi-rigid, integral PU foam and rear spoiler made of flexible, integral PU foam.

After the first driving trials with a front spoiler made of integral, semi-rigid PU foam (produced from a synthetic resin mould), the shape of the spoiler was optimised, its weight reduced and the method of fastening it laid down, fig. 4. After the initial foaming tests the required degree of shrinkage for the metal tool was likewise laid down. The degree of shrinkage especially must be determined as accurately as possible since a moulding of this size can vary by a few cm if the degree of shrinkage laid down is too generous.

Fig. 4. Fastening of the front spoiler to be bumper.

Not only the normal moulding shrinkage should be taken into account, i.e. shrinkage after demoulding, but also post-shrinkage which occurs during painting.

If the synthetic resin tool is treated with care, several hundred foam components can be made with it. Painting such components is, however, more difficult than parts made in a metal tool. Synthetic resin moulds are especially advantageous for special purpose vehicles or short runs. Not only are they very much less expensive but the parts can be made much more quickly compared with steel moulds.

Deciding on the foam system for the front spoiler for the BMW 635 CSi

When deciding on the right foam system for the front spoiler, the G modulus test was an important criterion of the experimental phase for assessing the system. This test enables the most important properties of a finished article made from integral, semi-rigid PU foam, fig. 5, to be determined. The G modulus figures obtained can, after some experience in this method of testing (G modulus test according to Gehmann), be transferred directly to practice.

Foam systems which show marked permanent deformation already after being subjected to the G modulus test, show very little resilience when subjected to wind and even if incorrectly stored, this foam system shows an

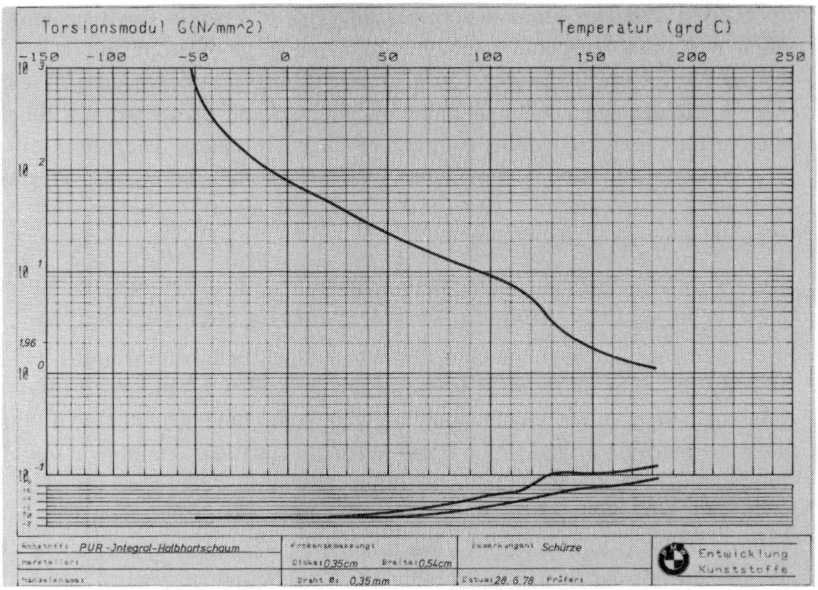

Fig. 5. G-modulus curve of the front spoiler foam system now being used.

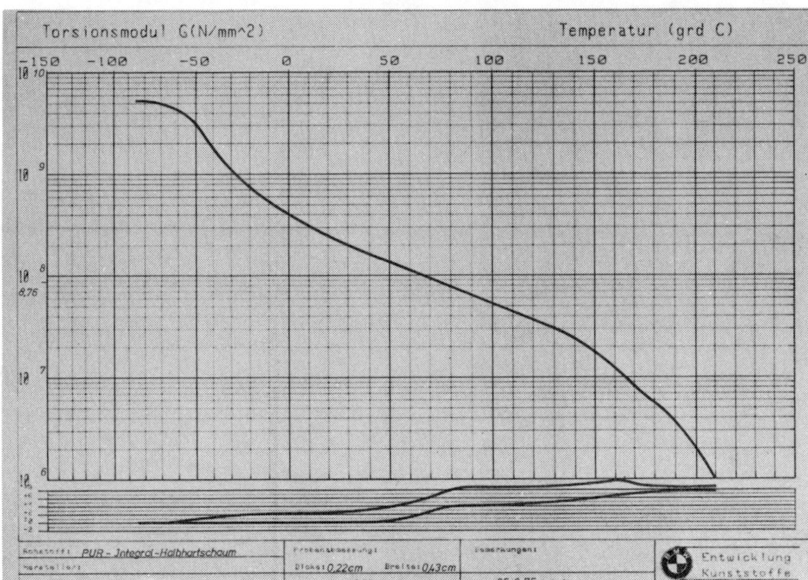

Fig. 6. Semi-rigid, integral PU foam system with big permanent deformation.

especially marked permanent deformation, fig. 6. Even in foam systems which, at normal temperatures, have a balanced range of properties, it is necessary to be on the look-out for permanent deformation because already at the painting stage a part may experience permanent deformation through being rested on an inadequate support, rendering it useless. The extent of permanent deformation of most foam systems becomes increasingly greater especially at temperatures above 50 °C so that appropriate remedial measures must be taken.

Deciding on the foam system for the rear spoiler for the BMW CSi

For the rear spoiler, too, the G modulus test provides useful information for selecting the right material. It is not only the rigidity of the material at different temperatures which can be derived from the G modulus curve, but the question of damping can be clarified for this relatively flexible material through a G modulus test, fig. 7. Another very useful test for an integral, flexible PU foam system for external applications is to examine microtome sections taken at right angles to the outer skin. This will quickly show whether the chosen foam system tends to absorb water or not.

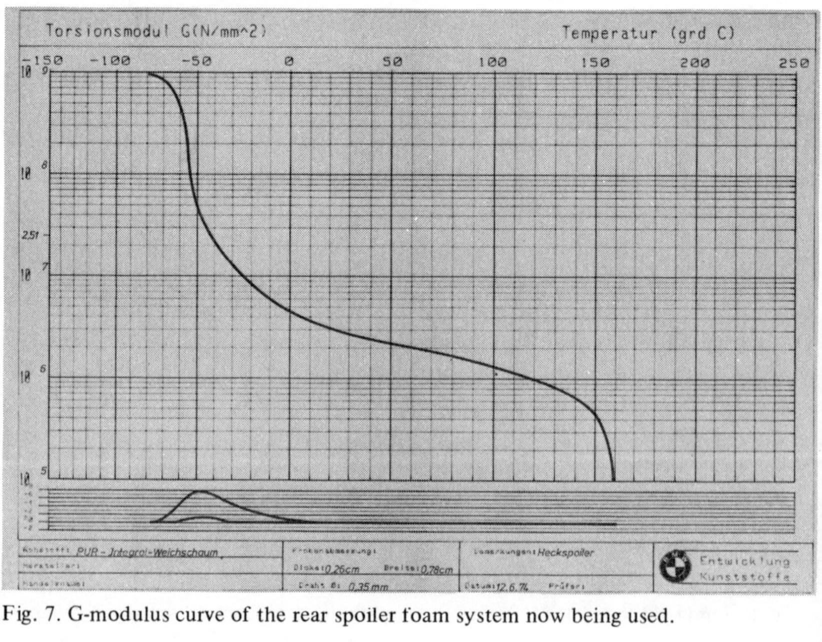

Fig. 7. G-modulus curve of the rear spoiler foam system now being used.

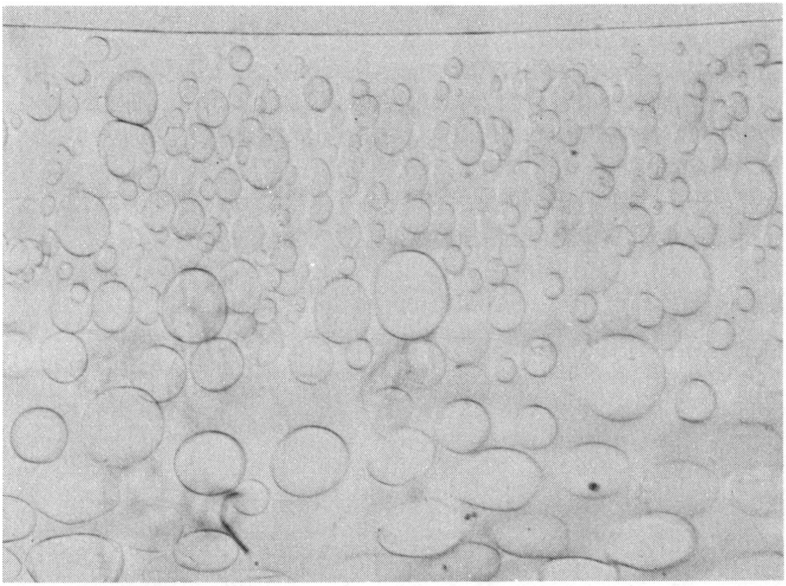

Fig. 8. Microtome section through a flexible, integral PU foam skin which does not tend to absorb water.

The outer skin of an integral, flexible PU foam rear spoiler should be as thick as possible and, above all, show a closed-cell structure, as shown in fig. 8. If the microtome section however shows that the skin is composed mainly of open cells which are physically linked with the core, as seen in fig. 9, an unpleasant surprise may be in store once winter comes, because the rear spoiler will become saturated with water due to capillary action and during the night the night frost will destroy the component. Not only the foam system itself but also the release agent has considerable influence on skin formation during the manufacture of a rear spoiler. If the release agent has been applied too thickly, or an unsuitable release agent has been used, tiny bubbles will be formed immediately below the integral foam skin, visible a first only under the microscope. The moment the part is painted however, these blisters will become visible on the surface.

Painting of front spoilers

Painting of semi-rigid, integral PU foam front spoilers still presents certain problems, even today. It is not a question of the paint itself but rather of

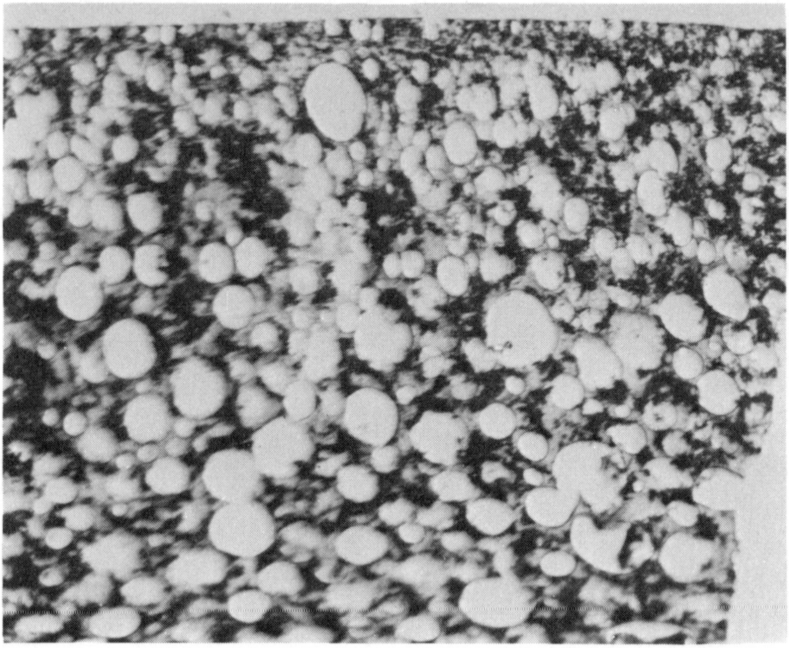

Fig. 9. Microtome section through a flexible, integral PU foam skin which shows a great tendency to absorb water.

the technique of application. Plastics components very quickly acquire a static charge and are very difficult to keep free from dust even through post-treatment. It is, moreover, absolutely essential to allow all the gas to escape from the mouldings after foaming has been completed so that blowing agent residues do not start to react when they are exposed to elevated temperatures during the painting process, causing bubbles to be formed on the surface.

Whereas, a few years ago, it was primarily one-component polyurethane paints that were being used to paint semi-rigid, integral PU foam components, which had to be stoved at about 120 °C, two-part polyurethane paints are today mainly being used in W. Germany, these requiring stoving temperatures of only about 80 °C.

Here, the special advantage is that spraying as well as overspraying can be carried out in car spraying shops. It is even possible to apply a standard car paint, together with a plasticiser, to a primed component. The physical properties of such a finish are still adequate for the applications in question.

Painting the front spoiler made of semi-rigid, integral PU foam to match the colour of the rest of the car no longer present an insoluble problem nowadays. The two colours can be matched with such accuracy that differences are hardly noticeable. If there is, for example, a black moulding to separate the plastics part from the metal carbody, even variations in shade are almost imperceptible.

The direct connection of a PU component to a metal one is much more of a problem, however. Here, the difference in shade in a new car may be viewed much less critically than the shade differences that occur between the metal and the plastics components as a result of different ageing of the various paints. In this critical assessment one must, however, be realistic enough to remember that even oversprayed metal parts in older vehicles show shade differences.

Seen overall however, the painting of a semi-rigid, integral PU foam component is still too complicated and thus also very cost intensive.

Painting of rear spoilers made of flexible, integral PU foam

The painting of flexible, integral PU foam components is extremely problematical, except black. Painting the flexible rear spoiler is not only desirable for visual reasons, but above all it helps to protect the part against weathering

Fig. 10. BMW 635 CSi with rear spoiler made of flexible, integral PU foam, with ornamental stripe.

influences. Unpainted PU rear spoilers have an orange-peel surface after about 6 to 8 months and, moreover, an unpainted surface additionally encourages water absorption which, in any case, is already critical.

It is not advisable to apply a succession of coat to flexible, integral PU foam parts because the resultant paint film will tend to wrinkle when the component is subjected to pressure so that, when the pressure is removed, the part will not completely recover its original shape. A wrinkled surface give the rear spoiler an unpleasant appearance.

There are, of course, isolated criticisms of black rear spoilers but even an ornamental stripe, screen-printed on to the PU paint film can add a touch of elegance to such a component so that it no longer looks like an alien appendage on the vehicle, fig. 10.

The development of polyurethane foams will, thanks to the experiences with front and rear spoilers, open up new fields of application for this material. The type of end use that springs to mind is large-area carbody components. The subject is already studied at BMW, in the form of the front wing painted the colour of the car and made from a glass fibre reinforced, semi-rigid, integral PU foam.

Self-supporting PU bumpers for the Peugeot 505

M. Basille

The function of the front and rear bumpers of a car is to protect the carbody from damage in case of a minor collosion. Although this has always been so, one must admit that chromium plated bumpers until very recently formed an important decorative element. The apparent strength and the well known properties of the materials chosen (stainless steel, chromium etc.) appeared to fulfil an insufficiently defined task.

The idea of "no damage" in case of accident enables one to obtain a clear picture of what is actually involved and to usefully compare various possible solutions. Evidently, this idea of "no damage" can be applied to definite parts of the vehicle – the bumpers alone, carbody plus bumpers etc., and it is here, particularly, that bumpers made of synthetic materials – to be discussed presently – can solve the problem in a way in which traditional materials are unable to do.

Stability and energy absorption

A normal steel bumper may be compared with a mechanically loadable bar, through its shape and cross-section (surface moment of inertia). This means that it is capable of absorbing great forces which, however, produce only slight elastic deformation. This applies to the entire length of the bar, although especially to the points where it is supported compared with the middle section.

From this it follows that this type of design, although capable of withstanding high mechanical loads, will prove unsuitable for absorbing the energy produced even in a minor collision, without itself being damaged.

Wherever a collision has to be cushioned without damage being caused, one should not speak of "stability" in a general sense, for impact may be defined by three factors: mass, velocity and type. The energy to be absorbed can be plotted by as many force/deformation diagrams as there are admissible deformations. The force/deformation diagram tells us the energy that can be absorbed by a given type of construction through integration of the area enclosed by the curve above the appropriate deformation.

Coming back to the classic bumper designs, personal experience has shown that these are not capable of giving protection against minor impact during parking, in traffic etc. without damage. They produce a loud noise, indicating locally limited impact causing major damage which bears no relation to the slow speed of impact.

The energy which a component must absorb with a very low elastic deformability, led to a very high loading peak at the end of the elastic region. Since the impact speed in this deformation stage has not yet been reduced to zero, the energy that still remains leads to plastic deformation, i.e. permanent damage of the bumper itself, detachment of the fastenings and damage to the carbody and chassis.

Let us now consider the other solution of the problem, with the aid of fig. 1. If the impact is absorbed by a bar made of another material, e.g. a plastic, corresponding to a low modulus of elasticity but considerable

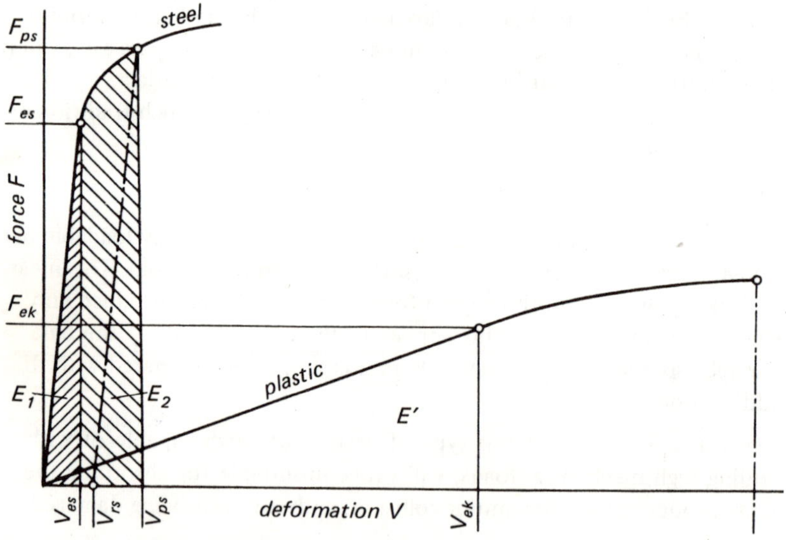

Fig. 1. Force/deformation curve for different bumper materials.

F_{ps} = load causing steel to be plastically deformed
F_{es} = elastic limit for steel
F_{ek} = elastic limit for plastic
V_{es} = elastic deformation of steel
V_{rs} = permanent deformation of steel
V_{ps} = elastic and plastic deformation of steel
V_{ek} = elastic deformation of plastic

elastic deformability, the impact energy $E' = E_1 + E_2$ can be absorbed with the very small force F_{ek} but with a greater elastic defomation V_{ek} i.e. without major damage (as well as without local damage for a larger area, as will be shown presently).

One can already see that a high modulus of elasticity of a material is no advantage – on the contrary, if one tries to keep the loading peaks (F_{ps}, F_{es}) which cause damage to the bumpers and surrounding area low, one must take into account the elastic absorption capacity in the case of such a problem of energy absorption:

$$E' = \frac{1}{2} F_{ek} \cdot V_{ek}$$

or:

$$F \infty E \frac{\Delta L}{L} \text{ and } \frac{\Delta L}{L} = \frac{\sigma}{E}$$

From this we obtain:

$$E' = \frac{1}{2} E \frac{\sigma^2}{2} = \frac{1}{2} \frac{\sigma^2}{E}$$

i.e. the quotient of the square of the permissible stress (generally the yield stress) and the corresponding elastic modulus which has to be taken into account, not the appropriate peak values and, above all not related peak values.

It is evident that the highest values of σ^2/E are encountered in certain groups of plastics, over a fairly wide range if one considers the ambient temperature as condition. The number of these groups is reduced considerably if one considers the extreme conditions – nowadays one considers only temperatures between – 30 and + 65 °C, as discussed below.

Consideration of weight

Where there is a sufficient, i.e. minimum distance between the bumper and the carbody, there would be a peak stress if this minimum distance were to be less, whatever the material used. This would inevitably lead to damage of the fastening or intermediate struts. Here, one will be able to limit the section thickness of the bumpers, thereby achieving a saving in weight compared with the classic solution, e.g.

Steelplate bumper:	section thickness 2.5 mm density 7.8 g/cm^3
Plastic bumper:	section thickness 6 mm density 1.0 g/cm^3

The weight ratio is $\frac{2.5 \times 7.8}{6 \times 1} = 3.25$ in favour of plastic.

It should also be remembered that, under these conditions, the plastics bumper, which is 3.25 times lighter compared with the one made of steel is also resistant to damage, the impact speed in both cases being limited to 4 km/h.

Resistance to local deformation through impact

In accordance with American and European specifications, a special profile (impact hammer) is used for the standardised simulation of a collision between moving masses (8 km/h and 4 km/h). This produces a load distribution of average intensity on the two extreme vehicle ends (extended length 406 mm, ends rounded, polished impact surfaces with wide angle).

Certain designers, analysing customers' complaints about the performance of conventional steel bumpers at slow impact speeds (1 km/h upwards) against small surface, such as a lamp post, have included in their conditions of acceptance the impact with an obstacle of 50 mm diameter (e.g. a post) at a speed of more than 2 km/h. The aim is to achieve dimensional stability at 4 km/h at every point of the bumper.

A condition which appears to be so harmless already results in a critical selection among plastics suitable for making bumpers that are self-supporting and therefore stronger. It was soon realised that thermoplastics with elastic moduli of 600 – 700 N/mm^2 and, incidentally, good resistance to deformation, are not suitable for bumpers of conventional cross-section, since they are damaged by 2 km/h collisions with a lamp post. The dent made by the lamp post is permanent – no matter what the supporting structure design.

We became convinced that only a "soft impact", compared with the unfavourable conditions of the "hard impact" would bring a solution in the case of a steel bumper (of the same energy). One need only look at the metal bumpers of a two year old car (or older) to see that its bumpers are dented along the entire length, the dents all being obviously due to similar causes. This is why one must go back to a material well known for its flexibility and resilience: polyurethane (PU). The materials known at that time (1977) were limited in their modulus of elasticity to $\geq 300\ N/mm^2$ and elongation at break of 200 %. Together with Bayer, we have now developed a new material which we have called HMU (high modulus urethane). The elastic modulus of this material had to be higher than $600\ N/mm^2$ with an elongation at break of more than 130 % at normal temperature, even under the most unfavourable conditions (position in the bumper, direction in which test specimen has been taken etc.).

We found that a bumper from one of the first experiments had the remarkable property of assuming its original shape after hard, local impact with a lamp post 500 mm in diameter at a speed of 3 and 4 km/h. Recovery of the original shape can, in extreme cases, take a little time, but is always less than half an hour.

One must be clear that this kind of behaviour differs fundamentally from that of the classic materials. It is rather like trying to make the impression of a Brinell ball disappear from metal – everybody knows that this simply won't happen.

In the case of HMU it was found that a locally very limited deformation, appearing plastic, very likely resulted in localised, considerable deformations under all combined experimental conditions (different tear speeds, bending, impact etc.) – but that this deformation disappears completely in an environment under elastic stress.

Variation of modulus of elasticity with temperature

Two extreme service temperatures apply to a bumper, namely – 30 °C and + 65 °C. Under normal conditions, bumpers made of HMU behave perfectly, being superior to ordinary thermoplastics. This superior performance is guaranteed 100 % through systematic post-treatment, after injection moulding, at a temperature of 120 °C. The elastic modulus at 65 °C is 45 % lower than at 20 °C ($660\ N/mm^2$ reducing to $370\ N/mm^2$).

$$\frac{E_{20°}}{E_{65°}} \approx 1.8$$

The material's self-supporting properties are guaranteed completely.

It is very important to take the perfomance of the material at – 30 °C into consideration, for at this temperature the recovery must be sufficient to ensure that, even with a hard impact, the bumper will remain intact. This break resistance is the result of increasing the elastic modulus (100 % whilst maintaining high recovery).

$$\frac{E_{-30°}}{E_{20°}} = \frac{1200}{660} \approx 1.8 \qquad \frac{E_{-30°}}{E_{65°}} \approx 3.2$$

At the same stress, the higher elastic modulus results in a lower deformation (≈ 0.5).

It is evident that the defomations that are permissible during fracture decrease much more slowly than those occurring in actual practice (see above). The result is an absolutely acceptable perfomance of HMU bumpers at – 30 °C, a temperature which is fatal for many polymers because it converts them into glass-like materials which tend towards brittle fracture.

Painting HMU

Bumpers are constantly getting bigger. Their colours are therefore becoming increasingly important to stylists, sales personnels and customers.

Bumpers which were more or less the same size as their chromium plated predecessors could safely be produced in natural colours such as black, grey and beige, also in the new shapes. This followed the trend towards "functional" bumper designs as opposed to the "ornamental" bumper. This trend was also encouraged by economic conditions.

As bumpers became bigger, making neutral colours rather monotonous, a desire for more cheerful colours naturally awakened. This applies especially where technically sophisticated designs exist and the unattractive characteristics of the unfinished bumper disappear. Here, HMU is the ideal material which can be painted without difficulty. There are paints on the market noted for the following qualities:

extreme resistance to wear, impact by flying road surface chippings and UV radiation

excellent adhesion to the substrate

identical expansion characteristics as the substrate.

To ensure these characteristics, it is sufficient to clean the unfinished bumpers by nomal techniques.

Here, it is important to point out that the paint will adhere satisfactorily to the HMU surface without any pretreatment. In other instances immediate adhesion is achieved by flame treatment of the substrate or application of a primer. As the painted object ages, and under extreme stress, one can see that this intermediate layer becomes detached from the substrate and subsequently forms wrinkles and finally drops off.

Conclusions

Car designers were given a new task: bumpers had to be developed which would completely protect the vehicle in case of a specific type of accident, i.e. they would then prevent any damage to the vehicle.

A material had therefore to be found which, above all, would have a favourable σ^2/E ratio and from which a protective component could be made whose size was in relation to the carbody and which would be self-supporting and in itself stable. Moreover, its shape should facilitate simple improvements compared with conventional bumpers.

The bumpers had, furthermore, to withstand very hard impact – greater than officially prescribed – such as is encountered when a car hits a lamp post. It should recover its original shape after impact and no brittle fracture should occur at – 30 °C, even in a serious accident. It should be possible to paint the bumper the same shade as the car body, without any of the above mentioned characteristics being lost.

Only the polyurethans will meet all these requirements. By adapting formulations we were able to produce PU grades whose modulus of elasticity exceeds 600 N/mm^2. This material we called HMU (high molecular polyurethane). With the help of HMU 670 N/mm^2 we were able to produce the bumpers for the Peugeot 505.

Trim and protective mouldings; ornamental stripes

Edmund Hellriegel

Never, during the various stages of development of vehicles, has there been any doubt about the question of ornamentation. The type of ornamentation and decoration was always adapted to the particular purpose. Chromium plated bumpers, window surrounds, strips to protect as well as decorate edges have now been replaced by simply designed structural units made up of various combinations of materials such as plastics, rubber, aluminium, steel and stainless steel. Function, economic considerations, weight, reliability and customers' wishes determine the extent of ornamentation on a car.

Trim mouldings, as shown in figs. 1 – 4, are used as follows.

To *protect edges* (EG Guideline 74/483, 485, 79/488), especially the top edges of the outer vehicle periphery.

To *cover* transition areas, overlapping seams and metal flanges.

To *mask* joints.

As *supports* with hidden surrounds for sealing strips.

Fig. 1. PVC/Tedlar strip.

Fig. 2. Ornamental stripes.

Fig. 3. Protective trim moulding.

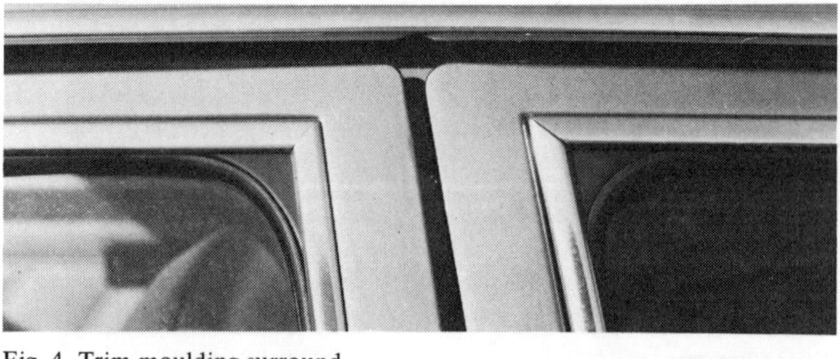

Fig. 4. Trim moulding surround.

To *mask* fastenings.

As a *surround* for front, side and rear windows.

As *decorative clamping strips* for window surrounds.

To *mark* the vehicle contours.

Design elements. Decorative profiles are used to achieve
instantly recognizable, showy or elegant design,
apparently longer and wider or elongated shape,
accentuation of character and shape,
emphasis of model variants through mouldings and emblems.

Decoration:

Metal trim mouldings and bonded Tedlar-aluminium foil combination.

Palladium applied by screen printing.

Screen prints, chromium plated.

Metallisation (chromium, stainless steel).

Pyrolytic separation.

Ceramic colours as required, diffusing into the glass surface during stoving.

Trim mouldings with impact protection have larger cross-sections. Impact energy is therefore more readily absorbed.

For these functions one generally uses metal-plastic or metal-rubber combinations, as shown in fig. 5. These external mouldings are mainly used where impact forces have to be absorbed and passed into load-bearing parts, fig. 6. At the same time, the aim is to protect the impact-sensitive outer skin as far as possible from possible deformation. The mouldings also protect other vehicles, e. g. when parking or opening doors.

Where the safety regulations according to FMVSS 214 have to be met, a rigid internal reinforcement is provided.

Trim mouldings as a protection against chippings

Loose chippings from road surfaces cause damage to paint films when they hit vehicles travelling at speed. Corrosion is the inevitable result and affects the value of the car. Those parts of the vehicle specially exposed to impact

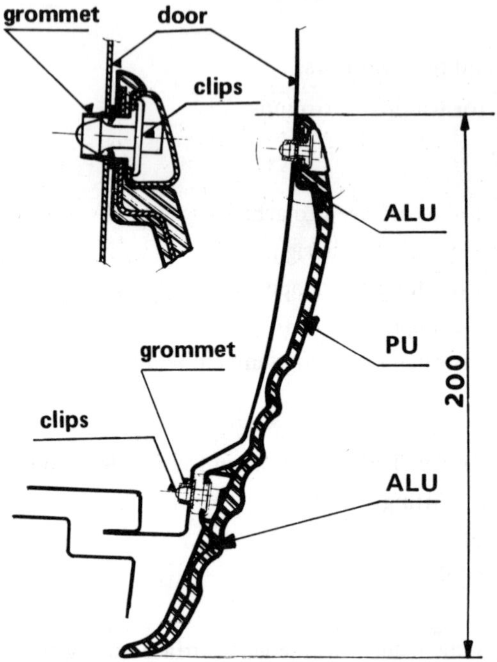

Fig. 5. Impact protection.

Fig. 6. Front ornamental and anti-impact strip.

by loose chippings, i. e. sills, wheel apertures and door edges are reinforced by masking strips etc. made of metal and plastics.

Wide, attractively designed decorative claddings protect entire side walls against external damage, dirt, water and dust. Damage caused by chippings

Fig. 7. Cladding to give protection against impact and stone chippings.

is prevented by the spray application of impact resistant paints. Trim mouldings mark the border between these protective coatings and the standard painted surface, fig. 7.

Stuck-on ornamental stripes to lend an attractive outer appearance to vehicles are now standard, ranging from broad bands to fine lines. These stripes are gaining in popularity over trim mouldings, figs. 8 and 9.

Fig. 8. Ornamental stripes.

Fig. 9. Ornamental stripes.

The development of single and double-sided adhesive tapes has played an important part in giving car bodies a colourful and attractive outer appearance. Standard vehicles can now be given a variety of designs, without adding to the weight, at little extra cost.

Whereas ornamental stripes used to be painted on laboriously by hand, they are now quite easily applied simply by sticking on the required number and colour combination of pre-printed stripes which are peeled off the backing film. The operation is carried out on the assembly line. These stripes are reliable and their adhesion and long-term performance are good.

Damaged ornamental stripes are easily repaired. Local defects can be repaired by sticking ornamental film over them, thus saving considerable paint spraying bills.

It is quite feasible that complete plastics outer parts are produced in a neutral basic colour, since ornamental stripes with different colours and design will produce the desired decorative effect and open up many possibilities for external decoration.

External plastics components combined with ornamental stripes can be produced with many different kinds of profiles, using struts, ribs etc. Plastics components are less susceptible to denting because they have a higher moment of inertia due to their greater section thickness. Because of the resilience of plastics materials, these resume their original shape after local deformation.

Coloured ornamental stripes stuck on to ribs and struts replace anti-impact mouldings. Thanks to the saving in material costs for ornamental stripes compared with trim mouldings, amounting to about 76 %, some of the extra cost of external plastics components is compensated.

Fig. 10 and 11 show a *PUR front wing*

with a stuck-on ornamental stripe on a moulded-on bead, and

with stuck-on, flexible trim moulding, each 18 mm wide.

The advantages are:

no fixing holes in the carbody,

no welding bolts,

increased protection against corrosion.

Fig. 10. PUR wing ornamental stripes.

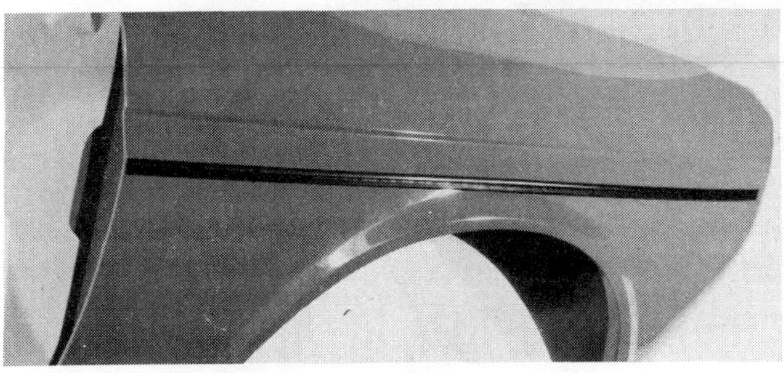

Fig. 11. PUR wing, PVC strip.

In the production of *steel carbodies* on the other hand, the bending operation must be taken into account, e. g. for doors, flaps and connecting parts, because continuous struts, ribs etc. greatly limit the use of pressing and bending.

Minor damage to steelplate bodies involves expensive repairs and re-spraying which is out of proportion to the actual effect of the damage on the vehicle's performance. Here, the flexible plastics outer skin with moulded-on impact absorbing bead is an advantage.

The criterion for the use of plastics external parts is still the cost of the materials and the effort and expense of achieving the same quality painted finish as on steelplate.

Development and production of trim mouldings

Plastics trim mouldings have been developed from metal and combined metal-non-metal mouldings.

Extrusion enables endless plastics strips and profiles to be produced in any required cross-section, with one hollow compartment or subdivided into a number of such compartments, with or without metal supports. It enables rolled metal profiles to be firmly joined to PVC, PUR, CAB and other plastics, figs. 12 – 15.

Fig. 12. PVC/aluminium extrusion, screwed.

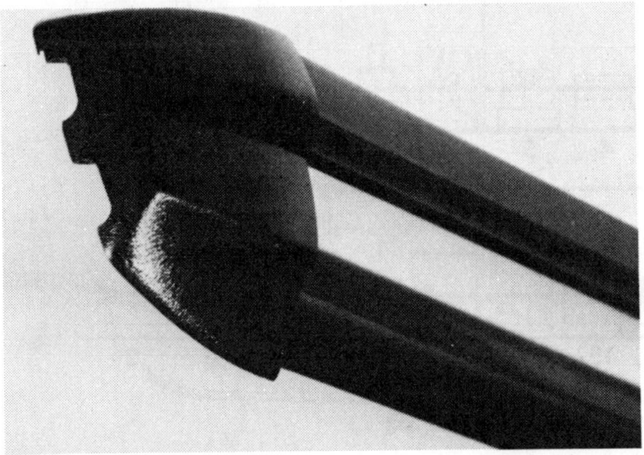

Fig. 13. PUR/stainless steel, bonded.

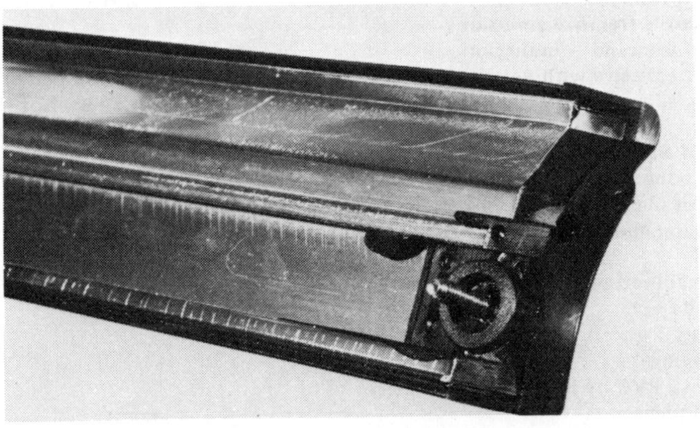

Fig. 14. PUR/aluminium, screwed and clipped.

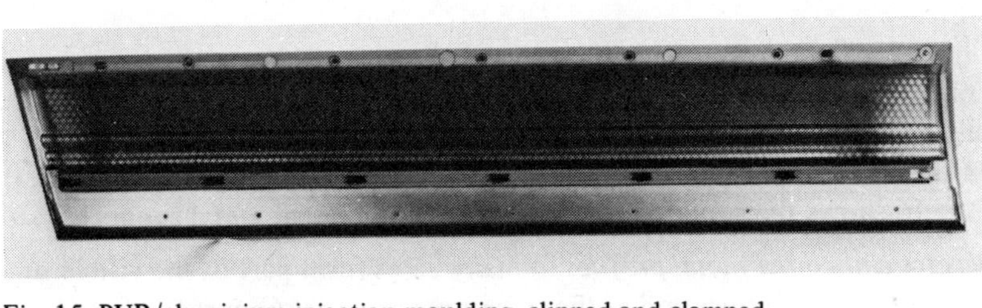

Fig. 15. PUR/aluminium injection moulding, clipped and clamped.

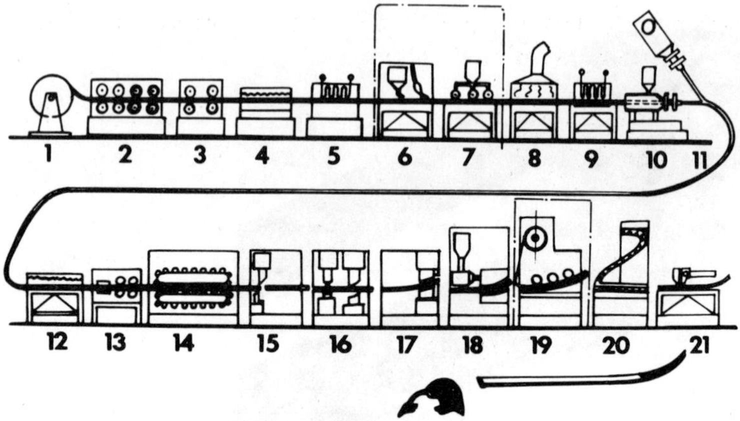

Fig. 16. Production of a trim moulding with metal support and moulded-on end piece.

1. Join coils endlessly (welding or bonding)
2. Profile the stainless steel/aluminium tape; 6 – 20 sets
3. Aligning: 2 – 4 rolls
4. Cleaning of profile (remove emulsion)
5. Drying of profile strand – induction
6. Application of adhesive with needle-type nozzle and spreading with felt-tip pen or
7. Application of adhesive with rollers, possibly coloured to give a check of even application
8. Evaporation of adhesive solvent
9. Activation of adhesive by induction loop
10. Co-extrusion of plasticised PVC with metal profile or
11. Parallel extrusion and bonding of plasticised PVC and metal profile
12. Cooling
13. Aligning and calibrating process
14. Take off profile
15. Cutting to size
16. Punching, shaping
17. Partially remove PVC by heating for end piece
18. Mould on end piece
19. Apply adhesive tape or film
20. Finish supporting elements (screws, clips, clamps)
21. Final control: surface, dimension, shape, position of supporting elements

Trim mouldings, with or without metal inserts, are environmentally safe and easy to repair. They cannot, however, meet the requirements of FMVSS 214 on their own – they must be provided with extra internal reinforcement. Fig. 16 shows the production of a trim moulding with metal supports.

Plastics embedded metal profiles in many cases form part of the visible surface of trim mouldings and at the same time serve as support for screwed or clipped-on models. Here they provide protection against mechanical and thermal stress as well as against shrinkage.

A good bond between metal and plastic presupposes adequate extrusion know-how. Heating may be by various systems such as:

conductive heating in the case of metal-plastic transverse strips,
inductive heating in the case of large-section metal-plastic profiles,
convective heating in the case of complicated external components.

Fig. 17 shows the price dominance of double-sided adhesive tapes among fastening elements.

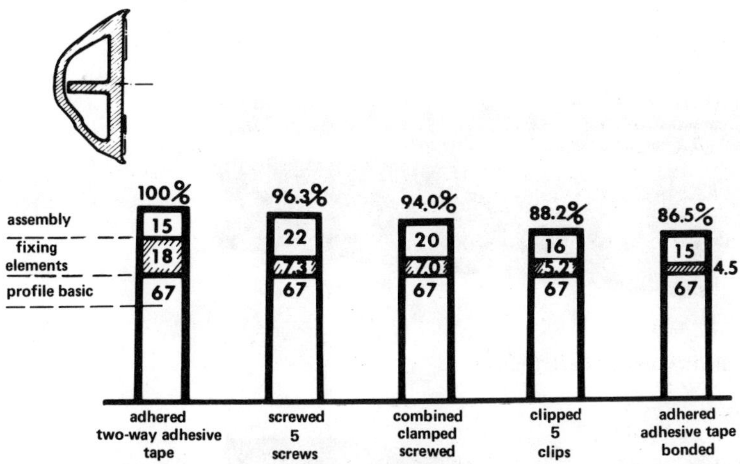

Fig. 17. Cost comparison of protective trim mouldings with comparable profile cross-section but different fastening elements.

If the advantages of bonded strips with narrow as well as broad cross-sections are to be utilised in future mass production, the bonding system will have to be simplified and its cost reduced. This can be achieved by rolling on the adhesive film last in the production of the strips and covering it with a silicone release paper. For the example in fig. 17 this gives the following proportion:

18.0 % cost proportion for double-sided adhesive tape
4.5 % cost proportion for adhesive film

and the following overall result:

100,0 % cost proportion for strips with double-sided adhesive tape

86.5 % cost proportion for the strip with adhesive film.

The latter is the least expensive of the five versions.

Foamed PUR protective trim mouldings are used in the following versions in vehicle construction:

with visible metal decorative strip, figs. 18 and 19

with a smooth surface

with Tedlar film

with PUR, without metal support.

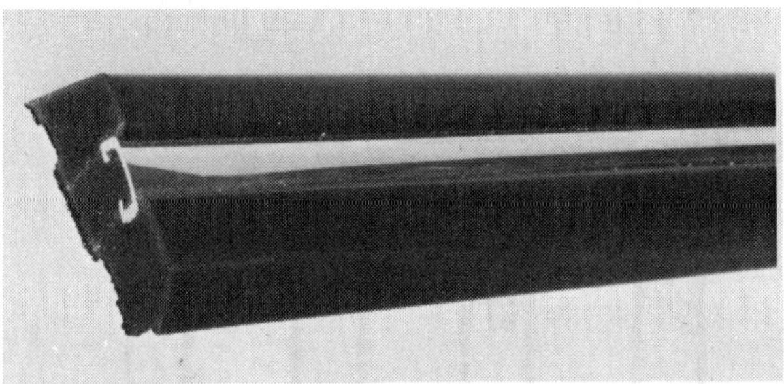

Fig. 18. PUR/stainless steel, foamed.

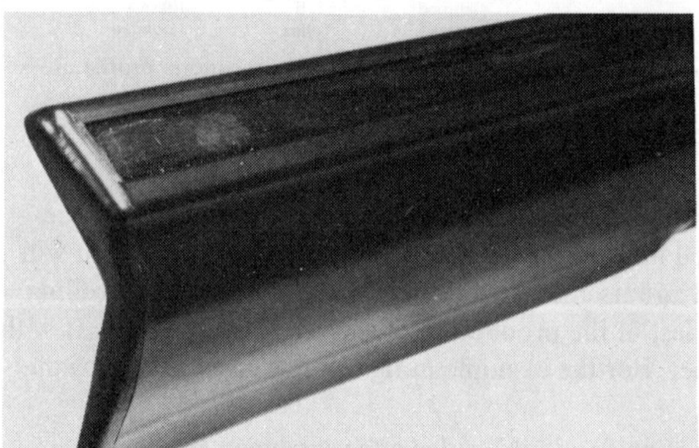

Fig. 19. PUR coloured/anodised aluminium.

PUR strips are injection moulded complete with end caps with metal inserts. The metal insert is perforated so as to obtain a better bond.

Fixing to the vehicle

bonding with PUR
bonding with butyl rubber adhesive
combined bonding with adhesive tape and adhesive
with adhesive film.

Multi-coloured, decorative and protective mouldings in which PVC acts as supporting material, are standard. The strips are made with or without metal insert by co-extrusion, figs. 20 and 21.

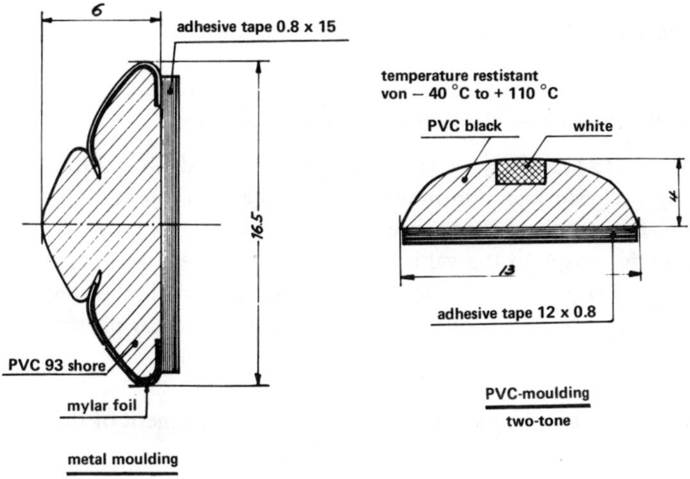

Fig. 20. PVC strips without metal support.

Plastics strips without metal tape inserts have the advantage of greater flexibility. They are better able to compensate surface tolerances than more rigid decorative trims with stabilising metal inserts, although the absence of a metal insert limits the width of the strips. They can also be stabilised by co-extruding them with one or more

plastics wires
plastics tapes or
ribbed metal wires.

Fig. 21. Aluminium/PVC with end cap.

Compared with metal tapes, this process is more expensive.

Large volume cavity profiles have caps moulded to the ends to seal them. With properly designed injection moulds, this process no longer presents any problems. One or two self-adhesive tapes are necessary to fasten the flexible protective strips to the vehicle, depending on their thickness. Tables 1 – 7 and figs. 22 – 32 show the construction, method of manufacture and application of various decorative profiles used for protective purposes.

The two component anti-impact strip, fig. 33, is a development of the future.

Characteristics

The carrier profile tape is extruded with one or more clip cross-sections and covered with adhesive film.

The anti-impact strip as second component is extruded or injection moulded and clipped on to the carrier profile tape on the vehicle. In the case of the injection moulded part, the end caps can be moulded on in such a way that they enclose folds or edges and prevent longitudinal movement.

The carrier profile tape is an improvement on fastening with clips or screws. It supplements the single or tape strip clip arrangement as a basis for anchoring strips, suitable for narrow as well as wide strips.

Table 1. Plastic/metal combination.

	Basis: extrusion
Technology	Extrusion of PVC over a metal support Metal support completely enclosed or with ornamental space
Design	Constant cross-section Plastics profile with or without hollow compartments Metal support as anchoring element, profiled for fastening elements Ends closed by moulded-on caps
Material	Stainless steel, galvanised steel, aluminium – enclosed in PVC – enclosed in PUR – possible with EPDM
Surface finish	PVC: smooth surface, largely without grain, black or coloured Stainless steel: polished
Method of fastening	Clamps, clips, screws Strips supplied ready for mounting Fastening elements moulded-in or fixed in position
Applications	Protection against impact and abrasion Sidewall strips, bumper covers, rain strips and edge covers, totally enclosed or with ornamental space
Criterion	Cross-section combination metal/plastic

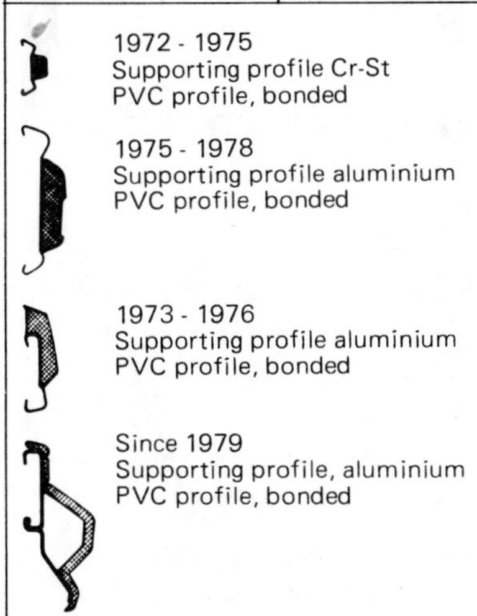

Fig. 22. Supporting profiles, metal/plastic, extruded.

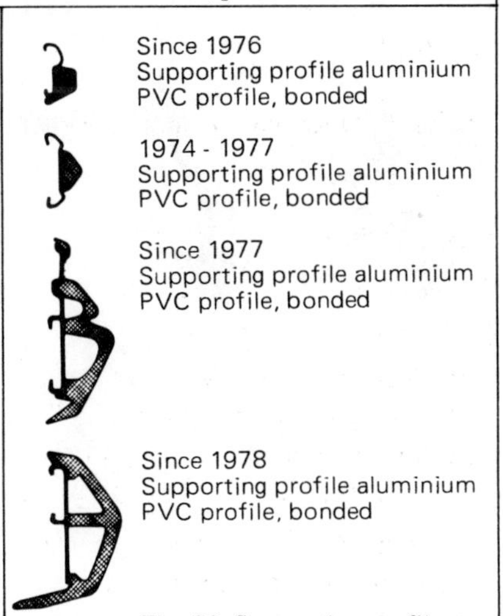

Fig. 23. Supporting profiles, metal/plastic, extruded.

Table 2. Plastic-metal combinations.

	Basis: stuck-on injection moulding
Technology	Separate production of the insert and the metal support by rolling or injection moulding. Separate production of the injection moulded part, followed by bonding
Design	Cross-section constant or variable as required End piece is not required since one-part plastics moulding
Material	Stainless steel, galvanised steel, aluminium or other material Plastics injection moulded part, bonded to metal components
Surface finish	Grained texture possible. Stainless steel polished. Aluminium can be anodised before or after bonding
Method of fastening	Clamps, clips, screws or with adhesive. Strip supplied ready for assembly. Fastening elements moulded-in or fixed in position. Adhesive tape applied during bonding
Applications	Protection against impact and abrasion, side-wall strips, window surrounds, bumper covers, channels
Criterion	Cross-section combination metal/plastic

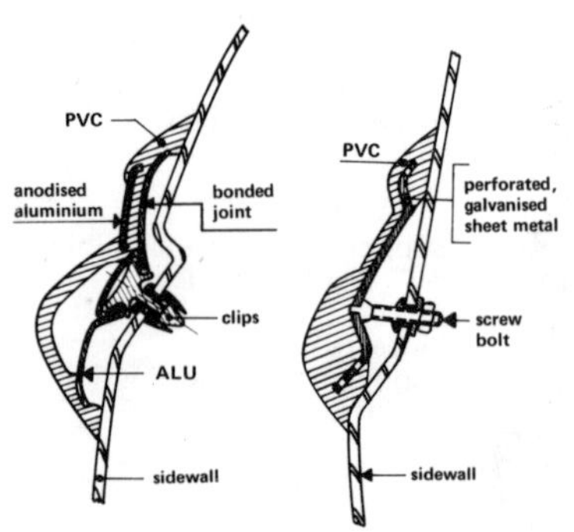

Fig. 24. Anti-impact strips, metal/plastic, bonded.

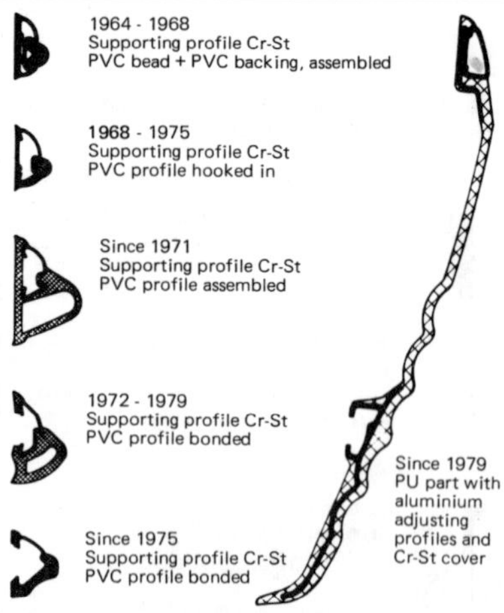

Fig. 25. Supporting profiles, metal/plastic, bonded.

Table 3. Metal profiles.

	Basis: rolled
Technology	Metal strip cold-rolled on profiling machines
Design	Constant cross-section, ends closed by caps or with plastics plugs, bent to shape (also three-dimensionally) with bevels, claddings, holes etc. Aluminium part components butt-welded
Material	Stainless steel, steel, aluminium
Surface finish	Stainless steel polished, aluminium glossy or anodised black, powder coated (PUR, EP). Steel slightly oiled
Method of fastening	Clamps, clips, screws. Strips bent ready for fitting, positioned with fasteners
Applications	Trim mouldings, ornamental trims, profiles to acco-modate window panes, rear window surrounds and windscreen frames, rain channel covers
Criterion	Ratio of cross section to bending radii

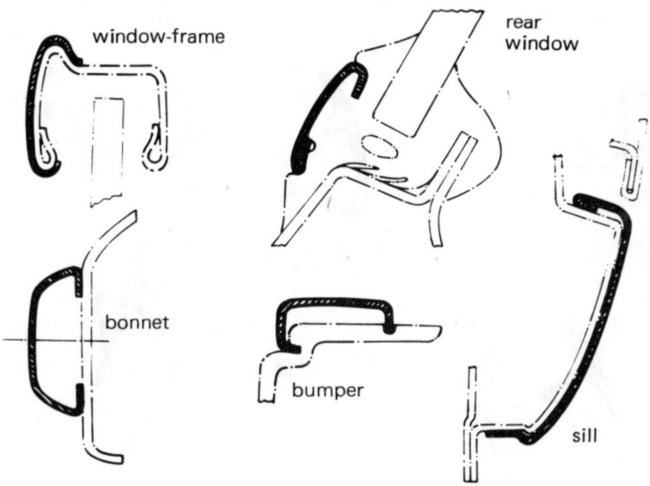

Fig. 26. Rolled metal strips.

Table 4. Metal profiles.

	Basis: extrusion
Technology	Extruded aluminium profiles
Design	Constant cross-section, ends cut, bent (also three-dimensionally) with bevels, holes etc. can also be made with plastics covering bead. Profiles joined by butt-welding.
Material	Aluminium
Surface finish	Aluminium, glossy or anodised matt-black or powder coated (PUR, EP)
Method of fastening	By pinning on, bonded into rubber profile clipped or screwed in
Applications	Trim mouldings, mainly as surround for side and rear windows and windscreens. As protection against impact with PVC covering bead
Criterion	Profile height to bending radius susceptible to scratching

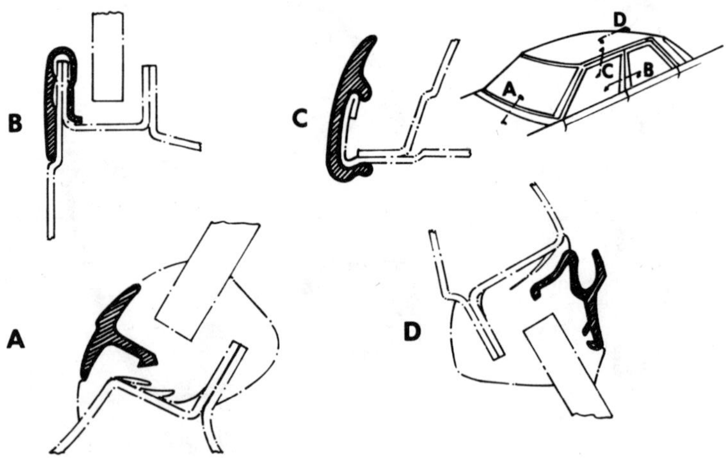

Fig. 27. Pressed metal profiles.

Table 5. Multi-purpose metal profiles.

	Window well sealing
Technology	Metal strip, cold rolled on profiling machines or extruded aluminium profiles with incorporated sealing profile, with plush or flock sprayed rubber profile
Design	Constant cross-section. Also available with incorporated sealing profile
Material	Stainless steel, steel, aluminium PP for sealing strips
Surface finish	Stainless steel: polished, aluminium: anodised glossy or matt-black, or powder coated
Method of fastening	Clamps, clips, screws, fixed ready for mounting, bent, with fasteners in position
Applications	Decorative and functional profiles for covering window wells
Criterion	Waterproof and airtight, with little friction

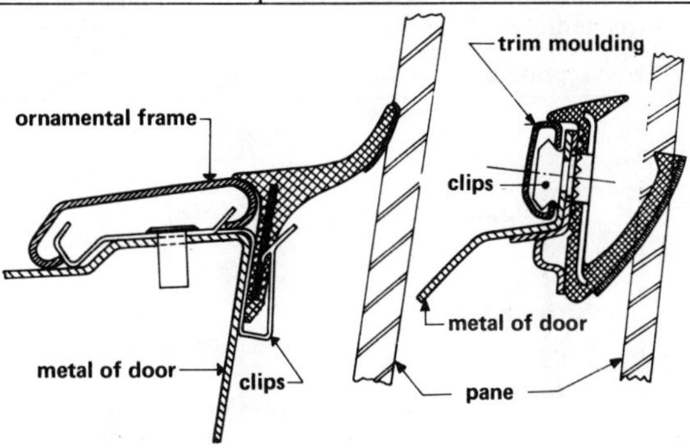

Fig. 28. Window well seal.

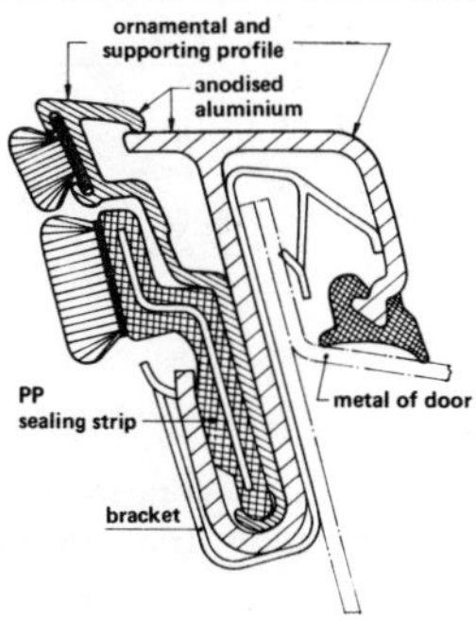

Fig. 29. Door/window well seal.

Table 6. PVC profiles.

	Basis: extrusion
Technology	Extrusion of plasticised PVC with or without light metal support. Coextrusion with transparent PVC outer layer over aluminium foil.
Design	Constant cross-section, supplied in roll form or cut-to-size lengths, with bevelled or moulded-on ends.
Material	Plasticised PVC with or without aluminium foil, for external applications aluminium is encapsulated, for internal applications on the outsides.
Surface finish	PVC transparent or coloured, ornamental space aluminium coated film oversprayed with crystal clear PVC with or without grain
Method of fastening	with self-adhesive film or with adhesive (e.g. Betasyl), PUR or butyl rubber. Profiled bead for welding interior trims (HF welding)
Applications	Adhesive strips for external and internal applications, ornamental profile as bead or HF welded on the film
Criterion	Cross-section and combination of various materials

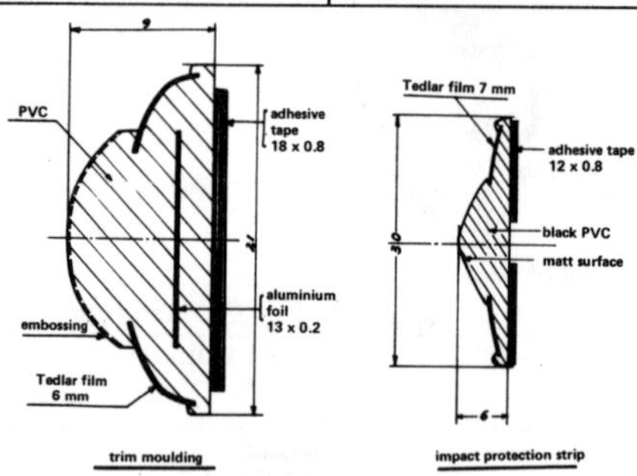

Fig. 30. Decorative and impact protection strip, extruded PVC.

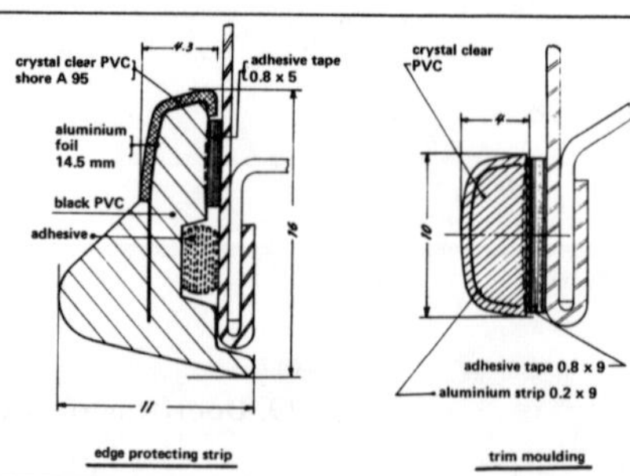

Fig. 31. Decorative and edge protection strip, extruded PVC.

Table 7. CAB profiles.

	Basis: extrusion
Technology	Extrusion of CAB with aluminium foil
Design	Constant cross-section, supplied in rolls form or in cut-to-size lengths
Material	CAB base support, transparent or with encapsulated aluminium foil
Surface finish	Transparent CAB with printed, embossed, brushed glossy or matt films
Method of fastening	By bonding, pinning, clamping, inserted as bead profile
Applications	Ornamental frames and strips, mounting profiles, trim mouldings e.g. rain channels
Criterion	Cross-section in relation to bending radius end deformation

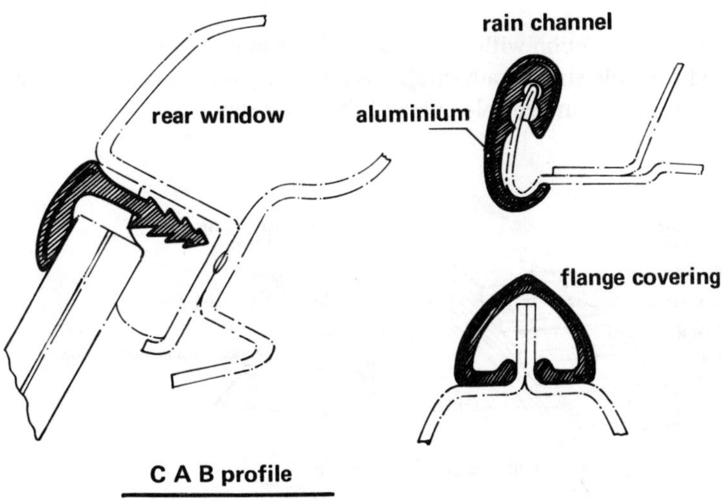

Fig. 32. CAB profiles, extruded.

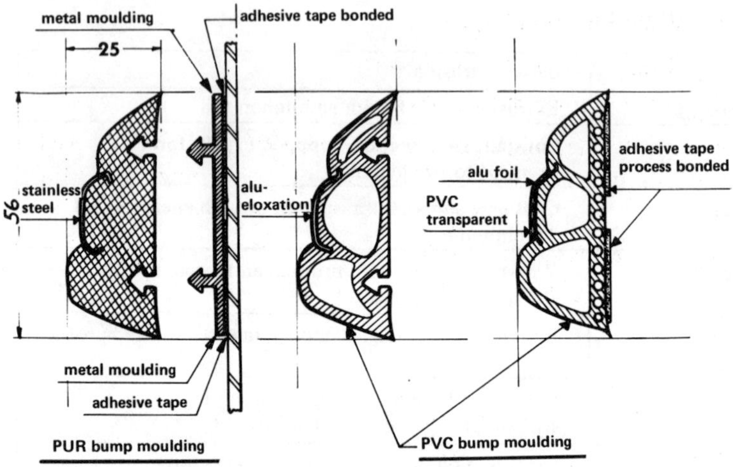

Fig. 33. Two-component anti-impact strip, extruded.

This new construction with an adhesive film is more economic. The bonding method for wide strips is advantageous especially for the carrier profile tape because it masks surface tolerances of the outer skin.

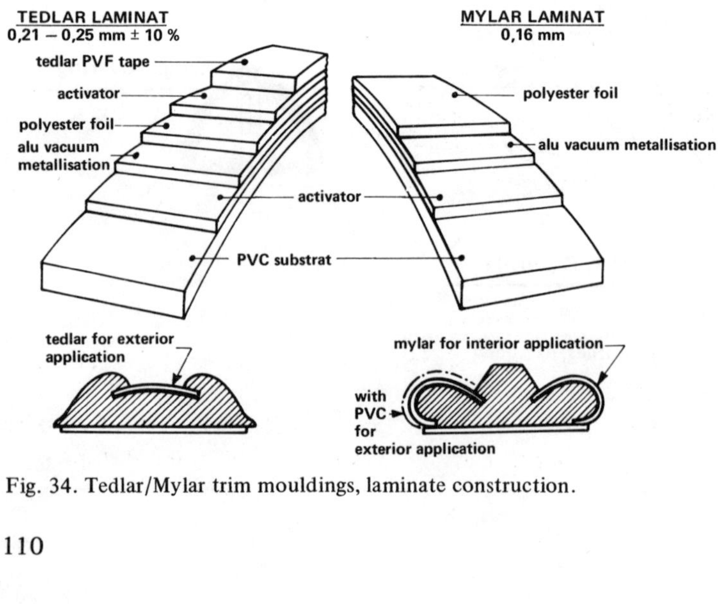

Fig. 34. Tedlar/Mylar trim mouldings, laminate construction.

Tedlar and Mylar trim mouldings, figs. 34 and 35.

Tedlar strips with sealed-on Tedlar laminate

Mylar strips with encapsulated Mylar laminate

Profiles made by this technique are identical in quality, provided the Tedlar film is flush with the cut edges that it protects the strip against ingress of moisture.

Mylar film must be sealed with an approx. 0.5 mm thick, transparent PVC film to ensure UV resistance. Tedlar film is UV resistant and can be sealed on the outside.

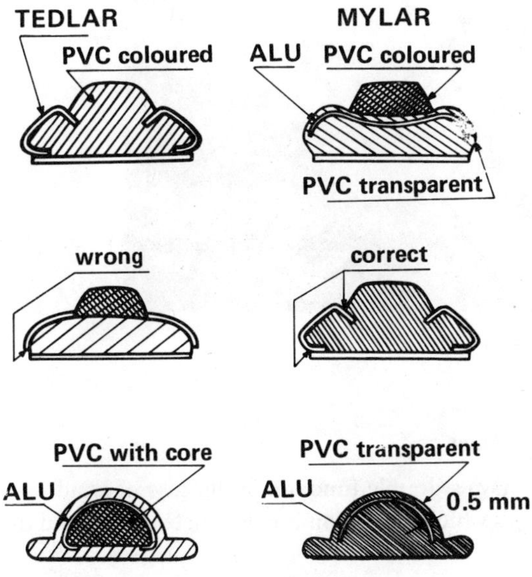

Fig. 35. Tedlar/mylar trim mouldings.

Mylar strips with and without cores

Strips *without* a core are made of high quality, transparent PVC and has a shrinkage of 0.8 – 1 %. There is little tendency to crease. Strips *with* cores have a shrinkage of about 2 %, depending on the PVC substrate of the core. There is a slightly greater tendency to crease than in the case of strips without cores.

Clamping strips

Mylar strip acts as sealant besides being decorative, especially for front, rear and side windows. It presses the two shoulders of the glazing profiles against the glass panes on the one hand and the carbody on the other, fig. 36. This clamping effect obviates the need to inject further sealant to make the windows watertight. The same effect is achieved in metal trim mouldings in place of Mylar windscreen and rear window profiles.

Fig. 36. Clamping strip, windscreen.

Trim mouldings with sealing effect

The following three profiles have a double function in the case of windscreens, rear and side windows which have been bonded instead of fitted in the normal way, fig. 37.

The trim moulding with sealant effect, as an extruded plastics profile with a Mylar or Tedlar surface is pressed into the gap between the glass pane and the carbody, fig. 38 left, when the panes are bonded to the body. It has two symmetrical sealing lips for glass panes and the edges in the body to accomodate the panes.

The same trim moulding, but with an asymmetrical sealing lip which is in contact with the glass, is shown in fig. 38 right. The other profile edge is in contact with the edge in the body which accomodates the pane.

Future methods of glazing in the car industry will involve bonding the trim moulding to the front face of the pane. The profile grasps the pane's edges

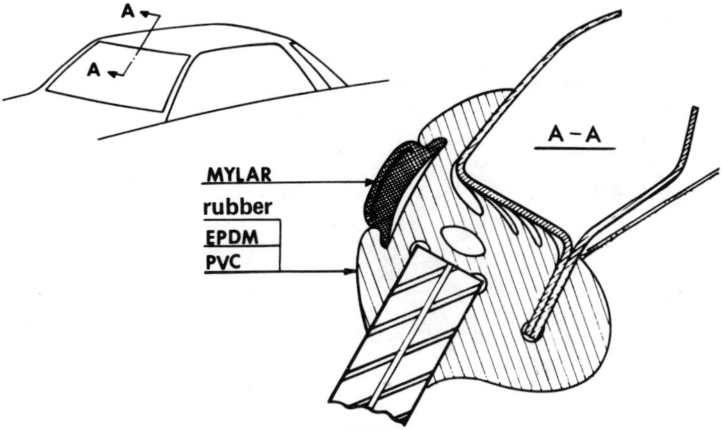

Fig. 37. Conventional front and rear glazing.

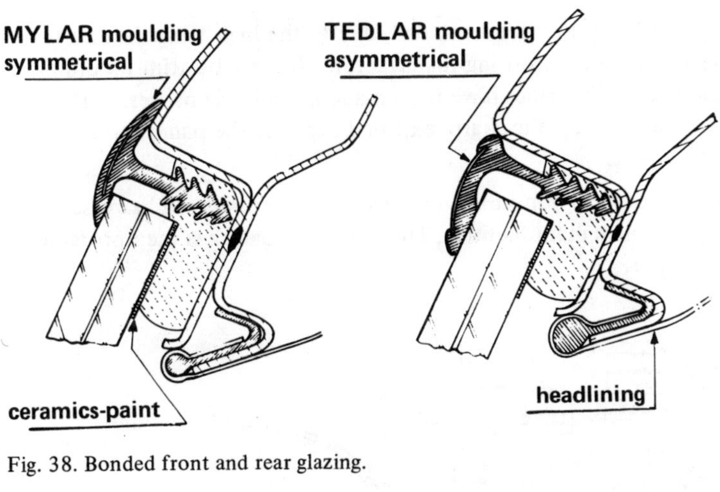

Fig. 38. Bonded front and rear glazing.

with two small lips. It is designed with an asymmetric sealing lip which covers the window opening edge.

Trim mouldings with a sealant effect are co-extruded. The visible surface, fig. 39a, is coated symmetrically or asymmetrically with aluminium or Tedlar laminate, stainless steel or aluminium profile tape. In basic models there is no lamination and the visible surface is natural-coloured. The plastics sealing lip and profile can be made in different Shore hardnesses.

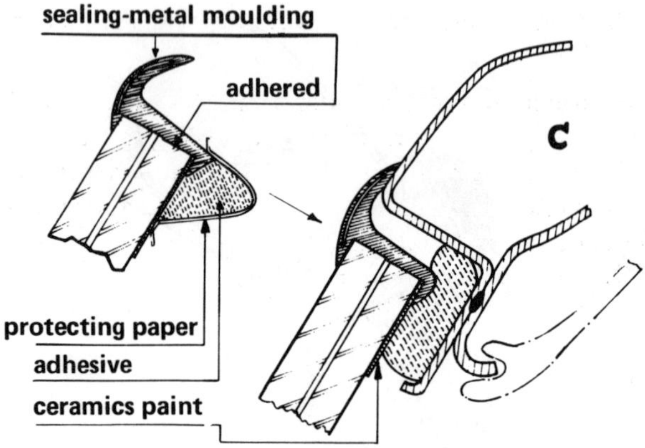

Fig. 39a. Front and rear glazing, strip adhesive coated.

Since the glass panes are free from dirt, due to the high temperatures to which they are exposed during manufacture, fitting the trim moulding immediately after the panes have been made is a reliable process which, at the same time, protects the sharp, exposed edges of the panes during transport.

The adhesives used to bond the windows to the carbody can likewise be applied by the glass manufacturer. The adhesive coating is then protected by release paper.

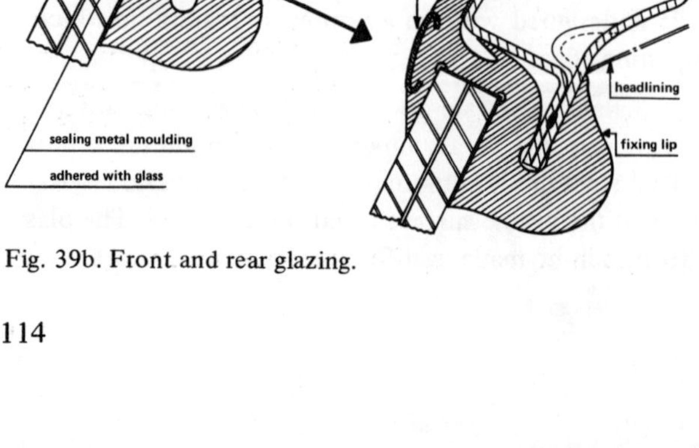

Fig. 39b. Front and rear glazing.

Another variant of the trim moulding with sealant effect has the following characteristics, fig. 39b.

The profile, with or without an outer ornamental strip is bonded to the front face and the inside edge of the pane. The outer sealing lip of the profile is covered, on the inner side facing the body, with an adhesive film which is protected by release paper. This lip is bonded to the aperture in the carbody for the window and exerts a sealing function. The inside sealing lip is drawn mechanically over the window aperture flange and exerts a supporting function. This variant offers advantages when windows have to be repaired.

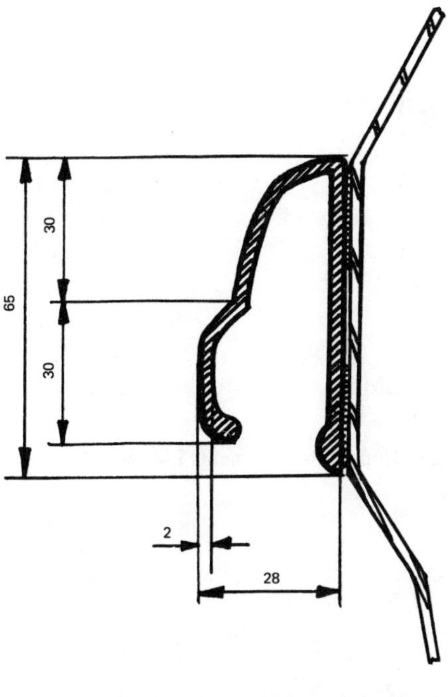

Fig. 40. Protective trim moulding, extruded EPCF profile.

A protective trim moulding of the future, fig. 40, is based on carbon fibre reinforced epoxy resin (EPCF) and may be extruded or press moulded with open or closed hollow cross-section. This would offer the following advantages over the present combined metal-plastic strips, provided that the price of carbon fibres were to drop to an acceptable level:

impact resilient

can be bonded with an adhesive film applied during manufacture profile caps moulded on to the open profile ends

EPCF protective strips are made in a standard colour or the visible part can be painted or covered with coloured adhesive film.

Weight saving 50 %.

In vehicles subject to impact requirements FMVSS 214, it is possible to design EPCF profiles as inside reinforcement similar to the present cross-section, resulting in a considerable reduction in weight, fig. 41.

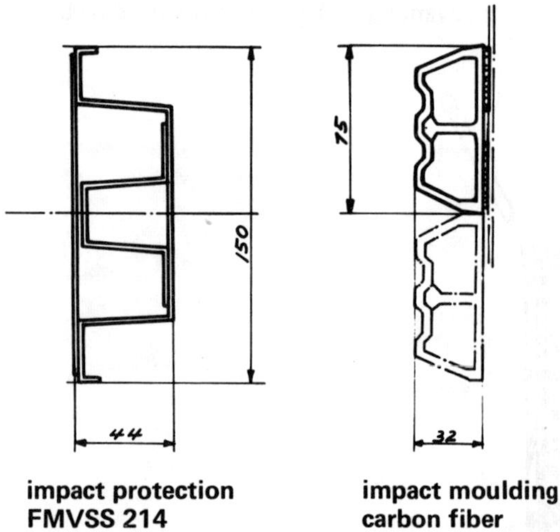

Fig. 41. Impact protection strip according to Impact Requirements FMVSS 214. left: metal, right: EPCF profile

Adhesive films, adhesive tapes

Bonding technique

Adhesion and cohesion are the fundamental conditions of any bonding operation.

Adhesion is based on the forces of attraction at the interfaces to be bonded. The shorter the distance between the surfaces to be bonded, the greater will be the forces of attraction. Surfaces which appear perfectly plane are, in fact, rough and fissured. Adhesives bridge these cleavages and fissures. The bond may be improved by applying pressure during bonding.

Cohesion is the holding together of the adhesive layer and is a measure for shear, tensile and bond strength, also under conditions of alternating stress.

Construction of adhesive tapes

Depending on the application, one can use unsupported adhesive films or double-sided adhesive tapes with different backing materials.

The adhesive

Requirements, fig. 42

high initial adhesion
high bond strength
temperature resistance continuous/short-term
chemical resistance
resistance to ageing, weathering, UV radiation, air, oxygen
surface finish.

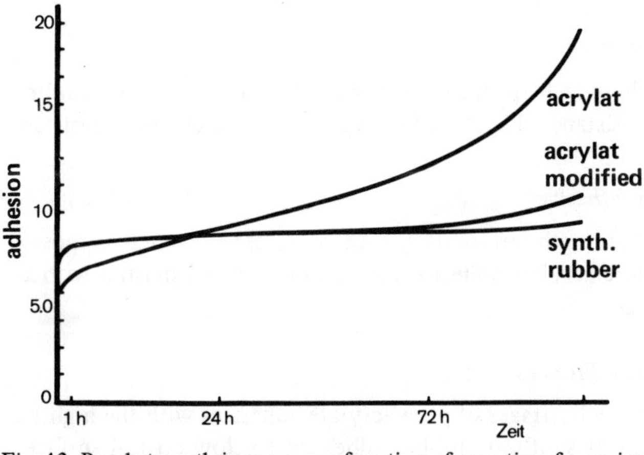

Fig. 42. Bond strength increase as a function of cure time for various adhesives.

Bonding to: painted surfaces, steel/aluminium, paper, wood, glass, ceramics, PVC. Plastics with high critical surface energy e. g. ABS, PVC, PMMA. Plastics with low critical surface energy, e. g. PE, PP, PS, rubber, fig. 43.

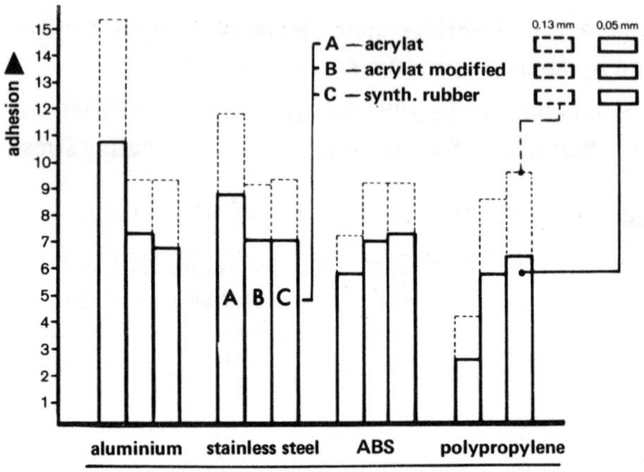

Fig. 43. Bond strength of adhesives to various materials.

Types of adhesives

The chemical composition and effectiveness of adhesives vary widely. There are four basic types of adhesive.

Pure acrylate adhesives

Pure acrylate adhesives form the basis for all adhesives with outstanding ageing and heat resistance, high bond strength and good chemical resistance.

Synthetic rubber adhesives

High initial bond strength but joints cannot be stressed as much as those produced with pure acrylate adhesives. Recommended for plastics with a low critical surface energy.

Modified acrylate adhesives

These combine the advantages of pure acrylate adhesives with the high initial bond strength of synthetic rubber adhesives for low critical surface energy.

Rubber-resin adhesives

These are suitable for general use and give joints with good bond strength. The initial bond strength does not increase. Moderate to good ageing, solvent and heat resistance. Bond strength can be increased by appropriate modification.

Types of adhesive tapes

Transfer adhesive tapes = dry transfer adhesive films without backing.

Adhesive	Thickness mm	Temperature °C	
		long-term	short-term
Acrylate	0.05 0.13 0.05	200 150 85	230 175 150
Modified acrylate	0.05 0.13	85 85	125 125
Synthetic rubber	0.13	65	80

Double-sided adhesive tapes = adhesive tapes with a backing which may be made of film or paper, both sides of which are coated with adhesive.

Double-sided adhesive tapes with foam backing.

Foam	Adhesive coating thickness		
	Acrylate	Modified acrylate	Synthetic rubber
PUR natural	0.8 – 6.4 mm		0.8 mm
PUR black	0.8 – 1.6 mm		1.6 mm
Flexible PUR foam, natural			1.2 mm
Neoprené black	0.8 mm		0.8 mm
PVC black	0.8 mm		
PVC white	3.2 mm		
Elastomer white	0.4 mm		
PP white	0.4 mm		0.4 mm

Glass fabric adhesive tape

High tear strength

Temperature resistance: – 55 to + 230 °C

Metal adhesive tapes

Systems acrylate, silicone, rubber
Temperature resistance: – 55 to + 300 °C

Nylon film

High strength
Heat resistance: 370 °C

The development of adhesive tapes has brought about the possibility of less complicated trim mouldings and convenient methods of fastening them. This also opened up the development of plastics mouldings without stabilising metal inserts.

Trim mouldings which can be stuck on are economic and with a calculable risk for shrinkage and long-term performance, provided materials are properly employed so that there is no risk of plasticiser migration. Depending on material specification, a plasticiser barrier can be dispensed with, see figs. 44 and 45.

Fig. 44. PUR anti-impact strip, bonded.

For *wide strips,* one can also use combined bonding to reduce adhesion risks through mechanical and weathering conditions. Fig. 46 shows two adhesive strips PUR or butyl rubber, paste-like, applied to the outer area of the strip, and a strip of double-sided adhesive tape in the centre.

To cure the adhesive one needs foam. The necessary air supply must therefore be guaranteed by an appropriate cross-section design because of the wax film applied to vehicles before they are delivered.

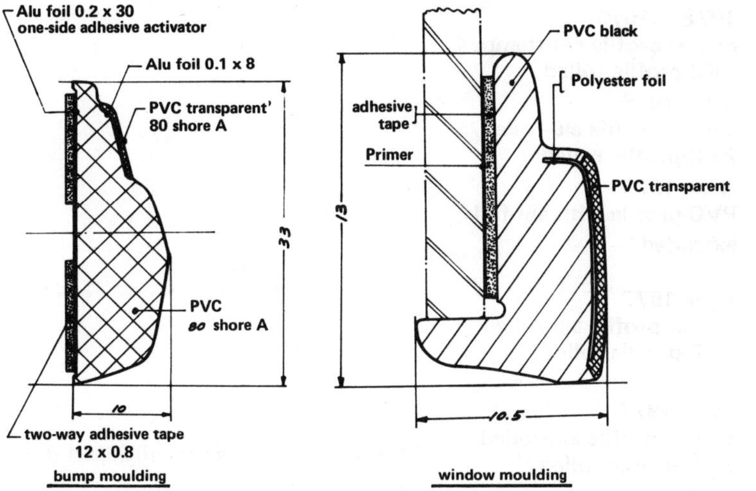

Fig. 45. Anti-impact strip (left) and window ornamentation strip (right) bonded.

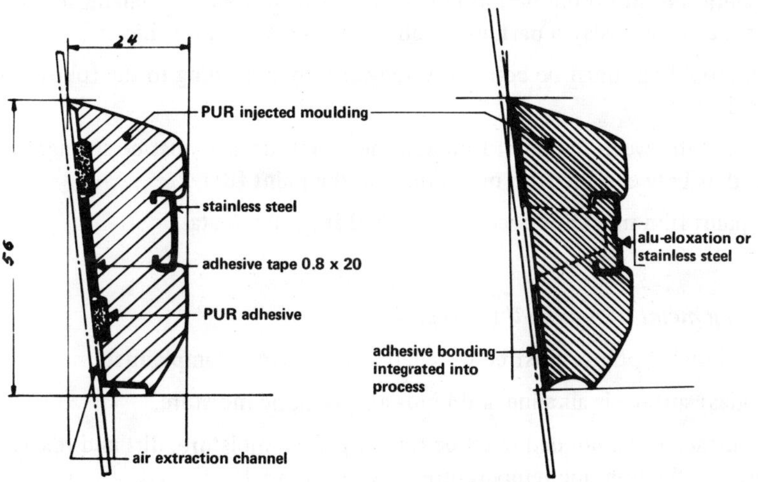

Fig. 46. Combined bonding for a PUR anti-impact strip.

The important point as far as bonding is concerned is the nature of the paint film. The behaviour of additives, especially in the case of high gloss paints, can adversely affect adhesion of the adhesive tapes to the paint film.

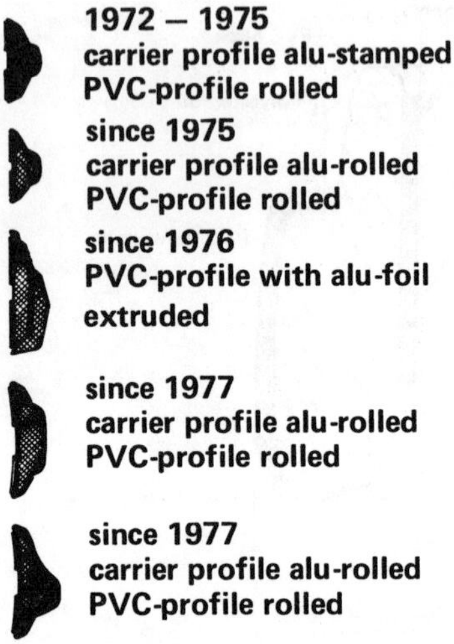

Fig. 47. Profile design of anti-impact strips.

Temperature, atmospheric humidity, chemical, physical, mechanical and dynamic stresses play a part, especially in the case of new vehicles.

Trim mouldings must be bonded to the carbody according to the following considerations:

The bond between the paint film and the metal surface must be stronger than that between the trim moulding and the paint film.

The paint film must not become detached from the metal.

Bonding metal trim mouldings to glass

The following points must be watched so as to avoid complaints:

The glass surface is alkaline and binds atmospheric moisture.

The surface to be bonded must be free of grease, moisture, dirt and release agent and the bonding temperature must be above 15 °C.

Etching or pickling of the surfaces to be bonded.

Correct choice and use of two-part silicone rubber adhesives.

Cyanate-acrylic adhesives.

PUR adhesives.

Metal profile edges to be sealed watertight with pane.

Bonding of combined Tedlar-aluminium film to glass

by means of an adhesive layer applied to decorative film (recommended)

with double-sided adhesive tape

bonding with adhesive paste

Screen-printed palladium coating, colour gold

(direction anodised aluminium), cannot be applied mechanically.

Method of production: the plane glass is screen-printed with an organic palladium compound before cutting and bending. This is stoved into the glass surface during the bending/pre-stressing operation, as a metallic coating.

Chemical and mechanical strength are excellent.

The abrasion resistance is moderate because of the softness of the material, so that this should, if possible, be applied only on the inside.

Screen printing, chromium plated, external appearance like chromium plated, trim moulding cannot be applied mechanically.

Production method: the plane cut glass is printed with a metal (silver) paste before bending and pre-stressing. This is then stoved into the surface during the pre-stressing and bending operation. The finished glass pane is galvanised, the surface of the frit being coated with chromium. Can only be used for the outside of the glass. Inside looks brownish/grey. Bending of glass edges possible in the case of certain printing techniques.

Properties: resistant against chemical attack and corrosion, very high abrasion resistance.

Metallisation: all corrosion resistant metals can be used, preferably chromium and stainless steel.

Colour according to metal used.

Production method: the bent, pre-stressed glass panes are metallised by high-vacuum techniques, areas which are not to be metallised are covered with masking tape.

Can be used internally and externally.

Chemical and mechanical resistance are excellent.

The abrasion resistance of precious metal coatings is very good.

Pyrolytic separation: Pyrolytic separation is used to deposit metal oxide coatings with great strength on to glass. The colour is similar to that of palladium, namely gold.

Production method: an organic metal compound is sprayed on to the hot glass surface, during bending/pre-stressing, in the form of a powder or solution. The compound decomposes on the glass surface and forms a metal oxide films with very good adhesion.

Chemical and mechanical resistance are excellent.

Parts of the glass not to be coated should be covered with masking tape.

Ceramic colours as required: This technique enables any colour to be applied to the glass surface. The inside as well as the outside can be coated.

Production method: the ready-cut glass pane is screen printed with a ceramic compound before bending and pre-stressing.

This is fired into the surface during pre-stressing.

The mechanical and chemical resistance are excellent.

A disadvantage is that the glass is lightly weaker where it has been printed. Ceramic colours can be used inside as well as outside.

Summary

Trim mouldings and protective profiles can be applied to nearly all types of vehicles today, thanks to highly developed process technology. Plastics are gaining in importance as competitive materials compared with the "all-metal" ornamentation formerly used. Development trends at the moment are centred on the combined metal-plastic trim moulding and anti-impact profile.

Realisable new developments

Two-component anti-impact strips with carrier profile tape, bonded, extruded or injection moulded strips, clipped on (easy to repair).

Inexpensive bonding thanks to adhesive film on metal-plastic strips.

Decorative and sealant strips on the front side of windscreen, rear and side windows, as functional gap covering of the bonded windows.

The foamed PUR injection moulding with metal tapes and smooth surface.

Flexible plastics outside with press-on impact bead profile covered with a bonded-on ornamental stripe.

Experiences with the Audi 100 with bonded windows

Hermann Stübig

In view of the ever decreasing stocks of crude oil, one of the most important objectives of the car industry is to try to cut down fuel consumption by making cars lighter. In the search for ways and means of achieving this, the direct bonding of fixed windows is assuming increasing importance because the resultant increased design latitude enables the c_w factor (drag factor) to be positively influenced compared with the normal method of installing windows with rubber seal strips.

The multi-component adhesive sealant compounds which were originally used, were replaced in the early seventies, after several years' development work, by a one-pack polyurethane adhesive sealant compound which greatly simplified mass production techniques as well as increasing strength.

At the Audi NSU works in Ingolstadt, work was started in 1974 to find out to what extent windows firmly and flexibly bonded to the car body could assume load-bearing functions in the successor to the Audi 100, these investigations centring on the windscreen, rear and side windows.

The following methods of bonding were known at that time:

Betaseal bonding
Butyl rubber strips
Solbitt method.

The decision not to make use of the butyl rubber strip and the Solbitt methods was taken for the following reasons.

The high contact pressure necessary to press down the strip evenly right round the window's circumference caused window breakages under mass production conditions. Moreover, the quality of the seal achieved was inferior to that obtained by the Betaseal process. For the CKD solution, with smaller numbers of vehicles, the use of butyl strips was admitted because of the limited shelf life of the Betaseal adhesive.

The Solbitt process requires a considerable amount of plant and equipment and thus does not permit a great deal of flexibility on the assembly line.

In deciding on the Betaseal system, a number of technical demands arose as regards the method of assembly used for the new Audi 100. Modern vehicle bodies should combine great rigidity with minimum weight. The firm, resilient bond between glass and body by means of polyurethane adhesive sealants offers the following structural advantages for the car:

In contrast to the use of butyl rubber strips, the adhesive, which cures to form a rubbery, resilient material, can absorb greater dynamic and static stresses acting on the vehicle body.

The technical data for the Betaseal adhesive are as follows:

Failure limit:	tension	$\sigma_B = 6.0 \text{ N/mm}^2$
	shear	$\tau_B = 3.3 \text{ N/mm}^2$

High and low temperature resistance: from – 60 °C to + 120 °C
Shore A hardness: approx. 45

The bonded windows comply with the requirements of the FMVSS test (Federal Motor Vehicle Safety Standards No. 212) already after 6 hours.

Because the glass is integrated as a load-bearing element, the vehicle body becomes more rigid. This means that certain components of the body can be made lighter. It has made it possible to achieve a very strong bond between glass and coachwork and the use of different shapes and sizes of windows mean that there is greater freedom as regards styling.

By bonding the glass to the body, up to 1.2 kg can be saved per meter of window circumference. The tolerances between the window and the body can be automatically equalised by means of the adhesive bead and breakage of the glass is prevented by the stress-free installation of the windows. Since better waterproofness is achieved, there is no need to patch up the seal once the customer has taken delivery of the car.

Since there were no comparable methods of installing windows being used on the European market and the bonding of windows on a mass production scale had not yet been tried, a new epoch of vehicle glazing was introduced.

The size of window openings in the car body which, unlike the conventional designs, were fitted with sealing frames, had to be considered in a new light.

In order to produce a big enough contact surface for the glass to be properly bonded to the car body flange, the flange width as well as the size of the window had to be balanced with respect to each other, as shown schematically in fig. 1.

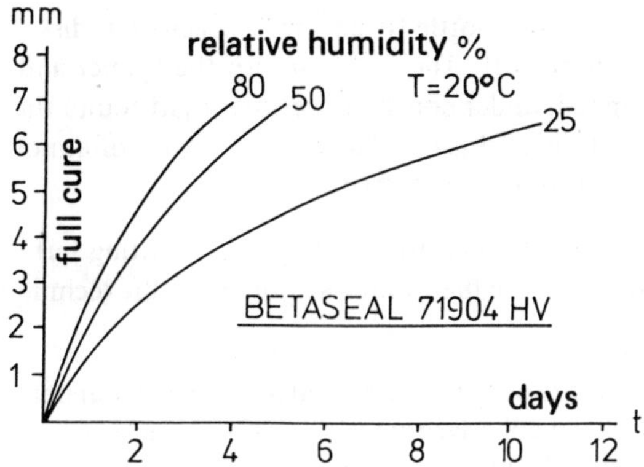

Fig. 1. Curing speed of Betaseal as a function of atmospheric humidity.

Since the car roof lining could no longer be bonded to the flange, a new method of fixing it had to be found. The solution consisted of integrating a U-shaped strip with the roof frame, which would enable the roof lining to be fixed by means of a profiled strip designed to also act as a barrier for the adhesive.

Since the sealing frame had now been omitted, another way had to be found for fixing the trim moulding. Tucker bolts were butt-welded to the window opening to enable clips to hold the trim moulding to be fixed.

To achieve good contact between the flange and the adhesive so as to achieve a tight seal, care had to be taken to ensure that the flange was absolutely plane – as is also necessary when using rubber sealing strip.

A number of operations were necessary, too, for this new system for pre-assembling the windows. The glass edges to be bonded must be cleaned with a mixture of isocyanates and silanes in organic solvents to remove all traces of dirt, grease etc. The cleaned surfaces must then be wetted with desalinated water and wiped dry.

To achieve a good bond between the cleaned glass edge and the adhesive, an approximately 20 mm wide strip of the glass is coated with a primer which has a similar composition to the cleaner mentioned above, the only difference being that it is pigmented black so that it also facilitates visual control. This black strip of primer also subsequently serves to protect the adhesive from UV rays, in conjunction with the trim moulding which is fitted later.

The primer is applied from a plastics bottle fitted with a special felt dispenser which must be free from fluff. The felt soaks up the primer and enables this to be evenly applied, under constant pressure, right round the window. The primer cures within 5 minutes, under the influence of atmospheric moisture and through evaporation of the solvent.

Attempts were made to obtain windows primed by screen printing but these were found to be more expensive than windows primed by the technique described above.

After the windows have been primed, they are coated with the Betaseal adhesive. One-pack polyurethane adhesives are high viscosity pastes when supplied, and are transported with the aid of a pump and applied to the glass with a specially designed metering device. Betaseal cures under the influence of atmospheric moisture – the greater the amount of moisture in the air the faster will the adhesive cure, fig. 1.

The adhesive is applied to the primed glass edge using a pistol equipped with a specially developed nozzle made of Teflon (PTFE), which enables the pistol to glide smoothly along the glass edge. Because of the special shape of the nozzle one obtains an M-shaped bead of adhesive which prevents the bead from collapsing if the atmospheric humidity is low so that curing takes longer.

The importance of applying the adhesive constantly along the entire circumference of the glass led to the development of a special pump with combined pneumatic-hydraulic control and a transmission ratio of 55 : 1.

A 200 litre drum of adhesive will be sufficient for bonding about 800 – 1 000 car windscreens or rear windows.

Since the adhesive will set in a damp atmosphere, it is necessary to use Teflon hoses in order to reduce the water vapour permability to a minimum. Furthermore, the windows must be in position within about 10 minutes of applying the adhesive.

Tests were carried out to try to apply the adhesive fully automatically. Because of the difficulties encountered, such as

uneven dimensions and planeness of the glass and

varying adhesive viscosities due to temperature fluctuations,

attempts to make use of automatic equipment were abandoned for the moment. Experience has shown that adhesive as well as primer can be applied just as accurately and quickly by hand by people who have received only brief training.

To ensure that the adhesive bonds firmly to the painted flanges, especially in the case of two-coat application, the edges of the window openings must be treated with a clear primer which resembles the glass primer in its chemical composition.

There had been some thought of using automatic equipment for installing the windows in the spaces provided. Technical considerations however prevented this being put into practice because the character of the joints made this impossible. Moreover, flexibility on the assembly line is maintained if the windows are put in by hand, because fully automatic equipment takes up a great deal of space in existing assembly lines.

Suction pads are applied to the pre-assembled glass to enable the windows to be handled. After putting distance pieces into the vent holes at the bottom, the window is placed into the opening, pressed in by hand and fixed in the correct position. The distance pieces in the vent holes prevent the glass from slipping downwards whilst the adhesive is setting, fig. 2.

The trim mouldings, developed specially for the bonded windows, have a special profile where they are clamped in position, rather like a pine-tree. When they are pressed down they engage the clips provided for the purpose. In this way the installation tolerances for fitting the mouldings to the upper edge of the roof can be equalised exactly.

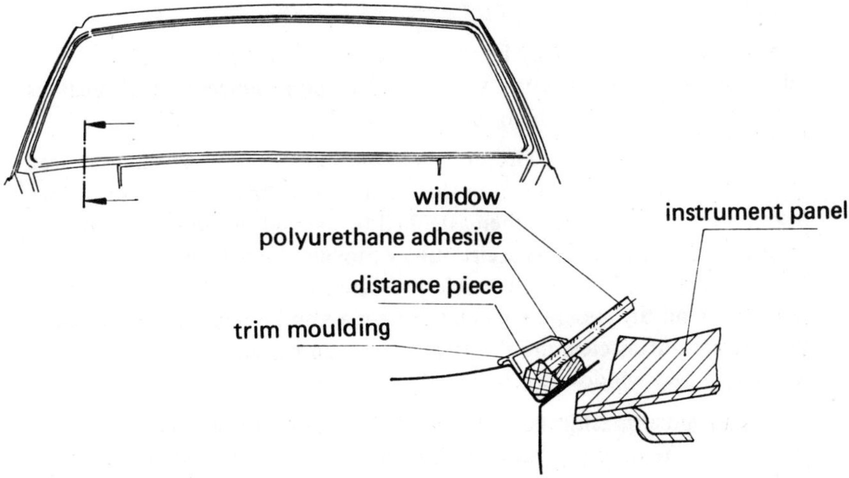

Fig. 2. Pre-assembled window.

The advantage of the black primer is that the trim moulding need not be in contact with the window. If there is a gap, this is not obviously apparent because of the black background. To prevent the trim moulding from being in contact with the glass and thus cause irritating noises at high speeds, a profiled sealing strip has been applied to the moulding.

This covers the entire assembly operation, fig. 3.

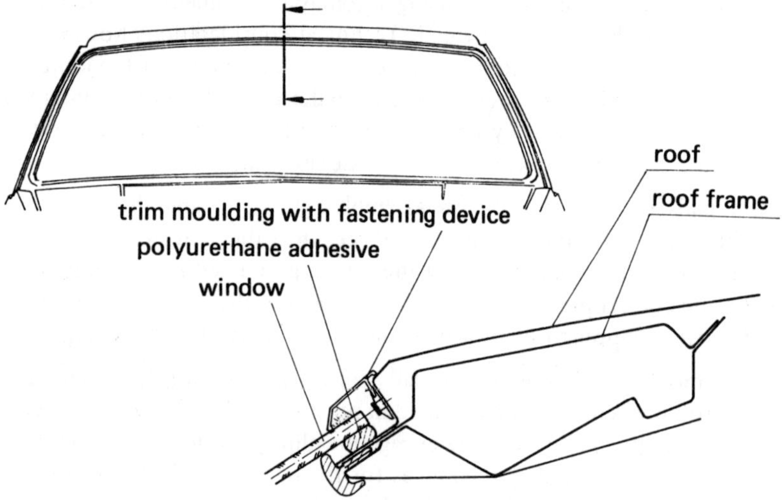

Fig. 3. Assembled window after being bonded to body.

The change-over to directly bonded windows also, of course, affects after-sales service. A special repair set has had to be produced, which enables windows to be easily exchanged, as described below.

A 0.7 mm diameter steel wire, to whose ends handles can be fixed, is used to pierce the adhesive bead from the outside, after the trim moulding has been removed. A handle is then attached to the end of the wire protruding on the inside, and the glass is cut out by moving the wire right round the circumference like a saw. To enable the wire to be moved smoothly and quickly round the glass, the gap between glass and flange should not be less than 4 mm. After cleaning the flange exposed through removal of the window, a new window can be fitted.

The improvements achieved with directly bonded windows, and the efficiency of the system, are certain to result in general acceptance of the new technique by the car industry. The finite element method and practical tests have shown that car bodies with rubber surrounded windows and those without windows lie in the scatter range of determinations for warpage of the vehicle body. An improved stress-strain performance can be expected when using Betaseal adhesive, fig. 4.

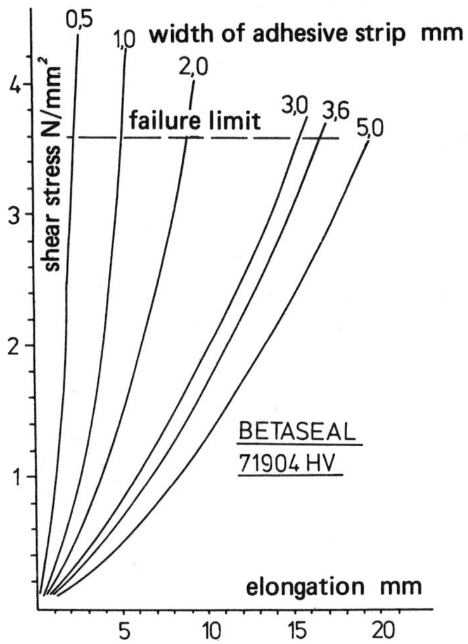

Fig. 4. Stress-elongation ratio for bonds made with Betaseal.

The separate parts used when bonding, e. g. car roof lining profile strips, Tucker bolts, clamps, trim mouldings with sound-proofing strips and the necessary preparation such as priming the glass and the vehicle body, make this method slightly more expensive than the fitting of rubber frames – but the weight saving that can thereby be achieved makes it well worthwhile.

Future models will continue to employ direct bonding of the fixed windows, since the introduction of some design changes, e. g. prefabricated car roof lining and plastics trim mouldings with special „pine-tree" profiles has meant that a number of components are now superfluous so that the cost ratio compared with rubber frames becomes more favourable and the reduction in weight comes to the fore as an unequivocal advantage.

Plastics headlamp lenses

Reinhard Röpke

There are four main reasons for wanting to replace car headlamp lenses made of glass by those made of plastic. These are: improved quality, design latitude, reduced costs and lighter weight. All the tests carried out with plastics lenses have shown that – as indeed with any modification of a vehicle component – it is the operation of the whole unit, in this case the headlamp, under various conditions of use, which must be taken into consideration.

The lens as an integral part of the headlamp, subject to suitability tests

Its primary function makes the headlamp lens just as much part of the headlamp as the bulb and the reflector. Fig. 1 shows the construction of

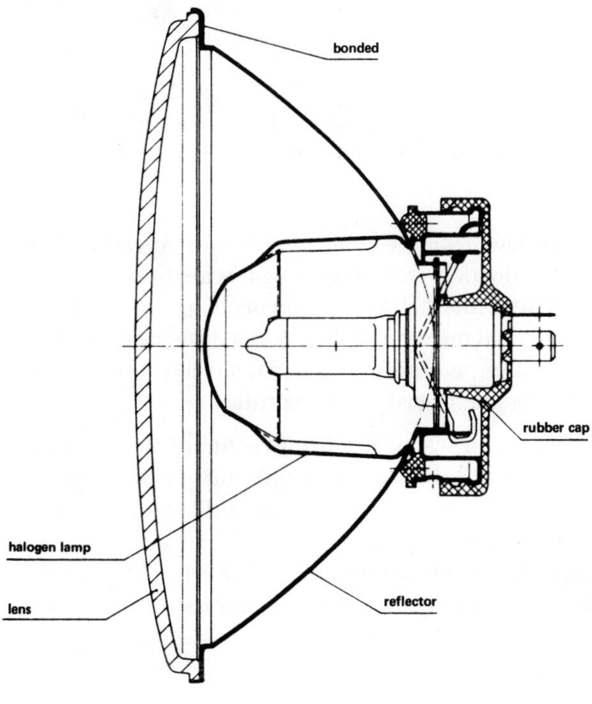

Fig. 1. Design principle for a headlamp unit.

a headlamp with halogen lamp, in which the lens and the reflector have been bonded together to form a sealed unit. To achieve a certain degree of light distribution on the road, the inner surface of the lens consists of mathematically calculated, accurately moulded prisms and cylindrical lenses. Fig. 2 shows a diagram of the optically effective zones on a headlamp lens and the different types of illumination they are responsible for.

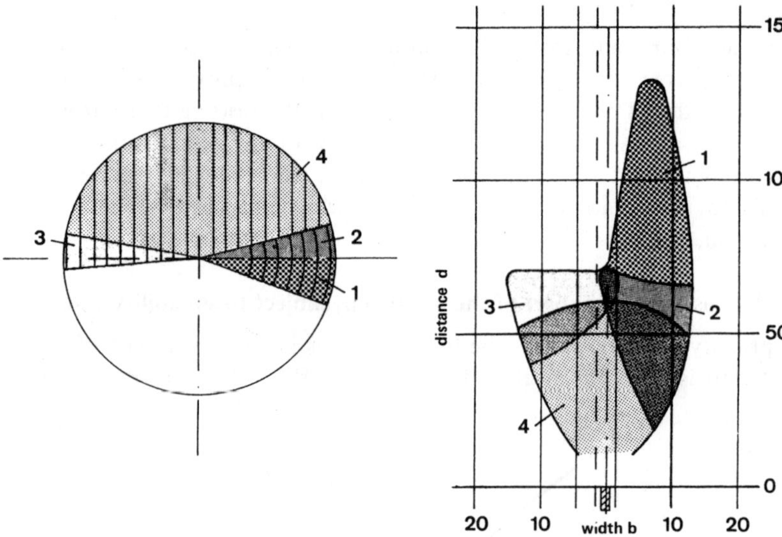

Fig. 2. Dipped beam – arrangement of zones on the lens and the roadway.

ECE and EEC regulations lay down the mode of operation of various kinds of vehicle headlamps. Illumination performance figures are obtained from the prescribed luminous intensities, minimum and maximum values for certain specified points and zones on a screen placed at a distance of 25 m in front of the headlamp. Authorised test institutes submit the headlamps to special tests prior to issuing the official quality certificate.

So far, European regulations for testing headlamps – unlike those in force in the USA – are almost exclusively concerned with their technical performance, in other words the photometry and the fixing of these units to the vehicle. The question of materials remained open, not least because 25 years ago, when the comprehensive European regulations for vehicle headlamps came into being, nobody even thought of plastics being used for headlamp lenses or reflectors.

For a number of reasons, glass and steel were the only conceivable materials in this field. Where, in the past, plastics had been used for such components in isolated instances, it was only because they had not been expressly forbidden. But in Germany, plastics can only be used in this field if this is expressly allowed.

To understand the problem, let us consider the basic functional differences between the ordinary lamps on a vehicle, whose lenses are naturally made from thermoplastic materials, and headlamps. The design of the former is determined by the fact that the emitted light need only be of low intensity and uniform, as long as it is visible. The function of a headlamp, however, demands a luminous intensity about 200 times greater, with considerable differences between the zones which have to receive the maximum amount of light and those which must only receive a minimum. The headlamp, in other words, is an essential aid to "active seeing". The beam must be accurately focussed and no light must be lost through scattering.

During the last two years, international committees which advise European governments in the drafting of regulations, have been working intensively on the problem of headlamp components, with special reference to plastics headlamp lenses. Even though this work has not yet been completed, it is evident that much can and must be done in this field. In addition to the photometric regulations, tests are necessary which simulate all the possible conditions to which plastics components may be subjected. Some of these tests will be done on the individual lenses, others on the headlamps as complete units.

When considering plastics for making headlamp lenses it is therefore necessary to take into account the practical function of these components, their relation to the carbody as a whole and the test specifications with which they must comply.

Design requirements for headlamp lenses

The optically effective parts of a headlamp lens must be made with great precision if they are to achieve maximum distribution of light. The angle tolerances required for the prisms in the asymmetrical zones (zones 1 and 2 of the lens shown in fig. 2) are ± 0.1°. This, together with the demand for sharply defined prism and lens elements and a high quality surface finish of the whole component, means that the part cannot be produced economically in pressed glass. This glass-related problem is eliminated by making use of injection moulding compounds and at the same time new possibilities are opened up through the use of plastics for headlamp lenses.

Plastics headlamp lenses also give stylists and carbody designers more scope as regards the shape of the headlamps. The maximum side ratio of 2 : 1 which applies to pressed glass lenses does not apply to injection moulded plastics lenses.

Chemical and physical requirements

Pressed glass headlamp lenses are almost completely unaffected by weathering influences. When using plastics however, one must ensure good UV resistance, i. e. guard against possible loss of light transmission. Lenses are subjected to Weatherometer tests lasting 2000 hours at 30 ± 5 °C with water impingement for 5 minutes, followed by 25 minutes without water, alternately. Polycarbonate lenses will have to be made from UV stabilised grades. Whereas there are no problems as regards the resistance of glass to chemicals and fuels, the resistance of plastics towards these substances must be ascertained by tests with the usual paint cleaners, tar removers and the like. Such products contain powerful solvents and, if applied with a dirty rag, strongly abrasive action is often also involved. Preliminary work on appropriate test specifications has not yet been completed. It is intended to use a mixture consisting of 15 % v/v diatomaceous earth, 10 % SiO_2 quartz W 3, 30 % n-heptane, 25 % xylene and 20 % toluene. The suggestion ist to allow this mixture to be applied to the test lens for 1 – 5 minutes, by rubbing it on with a force of 2 N cm^{-2}. The amount of light scattered by the lens after the test must not have increased by more than 4 %. The plastics chemist is thus faced with a daunting task.

Generally speaking, the dazzle factor is the most critical characteristic of a headlamp, especially if it has been designed to give a more intense light on the side nearest the roadway. Unlike glass, plastics lenses show greatly increased dazzle factors after being subjected to the following test, which can be described as follwos: test area: 14 x 14 mm; sand of rectangularity 1.8 – 2.0, Mohs hardness 7, particle size up to 0.2 mm; contact force 2 N cm^{-2}; 50 cycles with a ram speed $\leqslant$ 3 cm sec^{-1} for a 40 mm stroke. The amount of light scattered by the specimen should not increase by more than 4 %. It was found that plastics coated with a special lacquer must be used in order to achieve sufficient resistance to abrasion and the effects of fuels and chemicals. The coatings examined included "hard" films based on polysiloxane whose performance is similar to that of glass, and "soft" films based on standard binders such as are used in vehicle finishes to make use of their resilience. The best results were obtained with "soft" coatings on polycarbonate lenses but even these, with light scattering values of more than 6 %, did not meet requirements.

Headlamp lenses are inevitably exposed to high temperatures due to a variety of factors. A considerable percentage of headlamps has already been fitted to the vehicle when this is given the final coat of paint. The temperatures to which the headlamps are subjected during this operation are not the same everywhere. There are also differences due to the method of operation, e. g. the time taken up by breaks in production etc. These considerations should also include the methods of drying paint films in repair workshops where it may well be that the touched-up area is dried by infra-red heaters of great intensity, possibly quite close to a headlamp. It has been found that headlamp lenses may well be subjected to temperatures above 110 °C in the course of paint drying.

The bulbs fitted to headlamps, which at normal dynamo voltages of at least 13.2 V take up about 60 W when the lamp is dipped, develop considerable amounts of heat. Fig. 3 summarises the power inputs of different bulbs at different settings. The greater part of this power is converted into heat. Another factor which must not be overlooked is the dirt with which head-

Lamp	Test voltage	Power input
A lamp		
dipped beam	13.2 V	approx. 48 W
main beam	13.2 V	approx. 52 W
H4 lamp		
dipped beam	13.2 V	approx. 62 W
main beam	13.2 V	approx. 68 W
two-filament operation	13.2 V	approx. 130 W
H1 lamp	13.2 V	approx. 61 W
H3 lamp	13.2 V	approx. 61 W

Bild 3. Power input for lamps.

lamp lenses are, unfortunately, all too frequently spattered. The heating-up of the lens – whether made of glass or plastic – is much greater if the lens is dirty than in the case of the clean lenses tested in the laboratory. In any case, the amount of heat to which headlamp lenses are subjected during use depends on the shape and size of the headlamps. Inexpensive headlamps in which the lenses are bonded direct to the reflectors – especially the smaller models – result in overheating of the lenses. Larger headlamps with their own housings are more favourable in this respect. Fig. 4 shows the temperatures which were measured for the bonded, clan lens of a widely used, circular headlamp with an effective diameter of 182 mm, when using a dipped halogen H 4 lamp operating at 13.2 V and an ambient temperature of 23 °C.

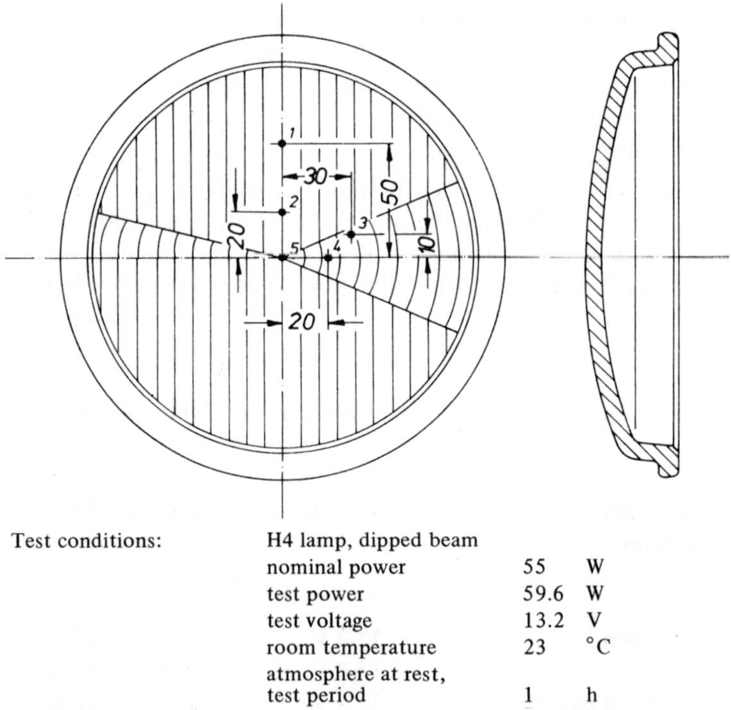

Test conditions:	H4 lamp, dipped beam		
	nominal power	55	W
	test power	59.6	W
	test voltage	13.2	V
	room temperature	23	°C
	atmosphere at rest, test period	1	h
Temperatures:	measuring point 1	105.5	°C
	measuring point 2	111.0	°C
	measuring point 3	115.1	°C
	measuring point 4	106.5	°C
	measuring point 5	78.0	°C

Fig. 4. Temperature determinations on a VW sealed beam unit.

Experience has shown that the temperatures encountered under these conditions do not present any problems for the hitherto used pressed glass lenses. Things are rather different however, when one is using thermoplastic materials. Even lenses made from thermoplastics will not immediately be destroyed by such critical temperatures, but there may be deformation which will have a very detrimental effect on the optical performance of the lens and may even critically change the headlamp's dazzle effect. The special problem is that the slight deformation which will be enough to alter the unit's performance will probably not be apparent by merely looking at it. Nonetheless, a headlamp which has been so affected can be an extremely serious traffic hazard. The test specifications soon to be introduced for plastics headlamp lenses thus provide for correspondingly stringent tests for heat distortion resistance. It is probable that headlamp lenses will have to withstand temperatures of 115 °C such as are encountered in car painting shops.

In addition, the headlamp ist operated for 6 hours with 15 minutes dipped beam and 5 minutes main beam alternately, with test voltage. This is a relatively harmless test, as long as both lamp filaments are not on during main beam operation. If necessary, the headlamp is tested accordingly and under these conditions plastics lenses will have to be excluded. It is extremely important, especially for smaller headlamps, to perform similar tests on dirty lenses. For these, the headlamp is covered with a sand/mud mixture of definite composition as specified in admissibility tests for headlamp cleaning equipment, which reduces the light intensity of the unit to 15 – 20 %.

After these high temperature tests, two defined measuring points on a screen 25 m in front of the headlamp must not show any change in luminous intensity. For the light/dark border a maximum displacement of 0.1 %, i. e. less than 1 cm per 10 m is permitted on the screen. These optical assessments are also provided for all other material tests to clear up doubtful cases. The demand for plastics headlamp lenses with high heat distortion temperatures will very probably make the use of materials with relatively high Vicat softening points, such as polycarbonate, necessary.

The use of plastics lenses will be especially advantageous where the lenses are subjected to considerable and sudden changes in temperature, e. g. when small headlamps which have become very hot are suddenly drenched in water. Such conditions could well be dangerous for normal, un-annealed pressed glass lenses.

The responsibility of manufacturers

General suitability tests prescribed for headlamps include individual tests on plastics lenses where necessary, some of which we have described above. It is generally understood that such specifications define minimum requirements. There is no doubt that headlamp lenses play an important part in ensuring safety. If they do not work properly they can endanger other road users. Manufacturers of headlamps and motor vehicles will have to decide where their own quality specifications are to go beyond the official ones or supplement them. For components such as headlamp lenses, the current European regulations – and those yet to come – will have to be taken into consideration. After all, who can risk putting a plastics headlamp lens on the market which could cause dangerous dazzle as the result of slight changes due to the effect of heat or chemicals.

Lighter weight

(The most important reason for considering plastics for making headlamp lenses is undoubtedly the ensuing reduction in weight, the actual extent of which will depend upon the size and shape of the lens. Weight savings of up to 60 % are possible, thanks to the lower specific weight of plastics moulding compounds (1.18 g/cm^3) compared with glass (2.5 – 2.6 g/cm^3) and the fact that plastics lenses can be made thinner, a 3 mm thickness being adequate.) It is, however, likely that the thickness of pressed glass lenses will also be further reduced. Fig. 5 summarises the weight of different kinds of headlamp lenses.

Vehicle	Weight per lens (glass 2.5 g/cm^3)	Weight per lens PMMA 1.18g/cm^3)	Possible weight saving per verhicle
VW GOLF	approx. 340 g each	approx. 136 g each	approx. 408 g per vehicle
AUDI 100	approx. 850 g each	approx. 340 g each	approx. 1020 g per vehicle
DB W 123	approx. 1500 g each	approx. 600 g each	approx. 1800 g per vehicle

Fig. 5. Possible weight savings resulting from the use of plastics headlamp lenses.

Cost comparison

It is not yet possible to give definite information about the relative cost of glass and plastics lenses. Besides acrylics, with their problem of insufficient heat distortion resistance, the only other material which could be used is the relatively expensive polycarbonate. The price development for the materials is more or less in proportion to that for glass and is a major factor in cost calculations. The injection moulding of these lenses is certainly no problem, but the necessity of applying a high quality coating to the lens surface, especially where hard coatings are required, means that a process has to be used which is by no means simple and requires a great deal of specialised equipment. The many problems – and their inevitable cost – make it necessary to use relatively expensive adhesives for producing the headlamp unit. In the production of pressed glass lenses on the other hand, considerable progress has been made in manufacturing techniques in recent years, especially where large numbers are involved. The cost of pressed glass lenses however is still relatively dependent upon their geometry. In general one can say that the moulds used for injection moulding plastics lenses are much more expensive than those used for making pressed glass ones. Glass undoubtedly is the less expensive material for smaller lenses with simple shapes, especially where these are made in large quantities. For small and medium runs of large, rectangular lenses, acrylics could offer price advantages. If, however, one must assume that acrylics have to be excluded because of their insufficient heat distortion resistance, the use of polycarbonate will mean considerably increased costs. A very high price has to be paid for the above mentioned advantages, especially for the weight saved.

Summary

Preliminary work, not only by producers of raw materials, but also by headlamp manufacturers and the experts who advise the authorities in drafting test specifications, has ensured that extensive information on plastics headlamp lenses is available. Nevertheless, further work and field tests are necessary, especially in view of the fact that, in 1980, the conditions for allowing such lenses to be used will be fixed. Fig. 6 shows where the various competing materials offer advantages. Plastics lenses seem desirable for a number of reasons, their main advantage being light weight and the precision with which they can be produced. On the other hand, one must not overlook the fact that such lenses do not necessarily exhibit the essential chemical resistance and resistance to mechanical influences. These drawbacks have to be eliminated if the use of plastics headlamp lenses is to be a viable proposition.

Assessment criteria	Glass	Plastic
Formation of optically effective areas		+
Design is in accordance with vehicle styling		+
Impact resistance		+
Susceptibility to scratching	+	
Chemical resistance	+	
Heat distortion resistance	+	
Resistance to alternating temperatures		+
Resistance to weathering influences	+	
Weight saving		+

Fig. 6. Comparison between headlamp lenses made of glass and plastic.

Automotive Plastic Composites

D. C. Travis and W. H. Fike

Accomplishments and Future Needs

The American automotive industry has made a serious commitment to use plastic as an engineering material in its cars and trucks. Considering the plastics usage by Ford Motor Company as representative of the U. S. industry, the content

Plastics usage per average Ford Motor Company Car

Model year	Weight (kg)
1960	10
1975	70
1980	103
1985	(Projected) 130 to 160

per average car increased from about 10 kilograms in 1960 to 70 kilograms in 1975.

The oil embargo, during the latter part of 1973, initiated a new era for the field of materials and process engineering. One of its major effects in the United States was the issuance of fuel economy standards for passenger cars, and more recently, for light trucks. Among the approaches pursued to achieve the fuel economy goal, vehicle weight reduction through the use of lightweight materials became, and still is, a major alternative. The energy-conservation and weight-reduction advantages of plastic over most other lightweight materials led to major programs for its implementation as a substitute for conventional steel. This resulted in an incremental usage of about 33 kilograms per average Ford Motor Company car between 1975 and 1980, to the current level of 103 kilograms.

Vehicle weight effect of lightweight material substitution

Material	Vehicle weight saved per kilogram used (kg)	
	Present	Future
High-Strength Steel	0.15	0.2
Plastics	1.0	0.7
Wrought Aluminium	1.0	1.1

With respect to future usage, our projections and plans call for the substitution of an additional 30 to 60 kilograms of plastics for heavier materials by 1985. It should be noted that, at the present time, the weighted car average of plastics substitutions represents a vehicle weight savings of one kilogram per kilogram of plastics used. However, in the future, this weight save factor is projected to decrease from 1.0 to 0.7 due to the implementation of components made with the higher-density composite materials.

To date, the growth in plastics usage can be associated primarily with decorative and functional applications in the interior and exterior of the vehicles, and components in the engine compartment and the electrical, fuel and climate control systems. Some of these components are noteworthy because they constitute an introduction of plastics to vehicle systems targeted for future applications:

Compression-molded, high-strength SMC bumper beams, which fit behind flexible urethane fascias or covers, were introduced in 1979 on the front and rear of four new Ford of Australia vehicles. A 60 % glass-reinforced polyester compound is used to mold the two components which are bonded with a structural adhesive to form the beam.

Low-profile SMC has been used successfully to produce compression-molded grille opening panels which can be painted with the car body, provide the desired surface appearance and fit to the body, and retain headlamp aim. Grille opening panels, for some 1980 models, are being injection-molded with a glass-reinforced polyester compound. A comparison of the two processes is in progress to establish relative benefits in terms of productivity and part quality and performance.

A bonded assembly of compression-molded SMC panels to produce a combination grille opening panel – radiator support – headlamp housing was introduced on the 1980 F-Series Medium Truck, Compression-molded SMC fenders are also new on this truck.

The use of flexible urethane fascias (end panels) has increased during the last few years. The 1979 Mustang front fascia, weighing 5 kilograms with an integral grille, demonstrates the capability of the reaction injection molding process and urethane material to produce large, complex parts.

Mineral-filled and glass-reinforced nylons are being used for several exterior body components such as, fender extension/headlamp housings, mirror housings, quarter extensions and lower back panels.

Functional applications which do not utilize composite materials have also been introduced recently in passenger cars:

Blow-molded, high-density polyethylene seat back/load floors for three-door vehicle models and station wagons introduced plastic to seating structures in 1979. This construction concept is being extended to similar applications in all car lines.

Blow-molded, high-density polyethylene fuel tanks were also introduced in the new Ford of Australia passenger car program in 1979.

Polypropylene splash shields and semi-structural fender aprons are being used on cars and trucks for corrosion protection.

It will become apparent that these initial applications of plastic to new vehicle systems serve as an intermediate step in the implementation of new technology.

Potential applications

Category	Typical applications
Exterior Body and Closure Panels	Hoods, Fenders, Doors
Seating Systems	Bench Seats
Structural and Functional Components	Frame Members, Wheels

To achieve the level of plastics targeted for future vehicle applications, large exterior body and closure panels, seating systems, and structural and functional components must now be converted to this family of materials. This means that programs must be actively pursued to develop composite materials, hybrid systems and component designs which will:

meet the more rigorous functional requirements of these new applications,

be compatible with current body assembly and finishing techniques, and,

assure high standards of product quality.

Also, to be competitive with alternative lightweight materials, every avenue must be explored to improve the productivity of plastic components, and to maintain their weight and cost effectiveness.

The first category of components which offers opportunities for new plastic applications is: exterior body and closure panels. The importance of these components to the growth in plastics usage requires that we evaluate every known process with potential for producing large panels, to identify:

the best material for surface appearance and product performance, and,

the most efficient process, as measured by variable cost and output of parts per dollar of investment.

Plastic exterior panels

Approach	Potential Application		
	Hood	Fenders	Decklid/ Liftgate
SMC – One-Piece	X	X	X
SMC – Metal Composite	X		X
SMC – Two-Piece	X		X
SMC with Molded Coating	X	X	X
Urethane RRIM	X	X	
Sandwich Molding			X
Vacuum Forming	X	X	X
Sheet Stamping	X		X

Significant progress has already been made in this area with the development of new polyester compounds and compression molding techniques. This technology has resulted in the capability to produce a one-piece hood with a good surface appearance, even though it incorporates reinforcing ribs on the inner surface. The combination of part design, low-shrinkage resin, multi-layered charge pattern with long and short glass fibers, and molding pressure staging results in a significant reduction of sink on the surface opposite to the reinforcing ribs.

More development is needed to extend this technology to larger flat hoods, and to other body and closure panels which have different structural and functional requirements. As interim approaches, the application of a functional coating on the surface of a formed and partially-cured part, in the mold, offers potential for eliminating porosity defects and reducing sink marks. There is also the hybrid assembly of a thin-gage metal outer skin with a plastic inner reinforcement. This concept takes advantage of the high modulus and the surface appearance of the sheet metal, and the lightweight of the reinforcing inner structure.

For other closure panels, such as doors, two-piece, bonded assemblies provide the best initial approach to achieve rigidity and exterior appearance, and to accommodate moving-glass mechanisms and attachment hardware. Multi-piece assemblies are also likely for Light Truck tailgates and station wagon tailgates since the inner panel in these components is used as a functional loading platform. Liftgates and decklids could possibly be one-piece components utilizing high-strength, low-profile materials. However, two-piece, bonded assemblies have to be considered now, to satisfy the structural and sealing requirements. The elimination of incremental facilities investment and assembly costs required for multi-piece components is a major incentive for the development of improved materials.

In the area of fixed body panels, fenders are candidates for plastics substitution. We are now producing compression-molded SMC fenders for medium trucks, and we are developing a passenger car fender with reinforced urethane materials processed by reaction injection molding.

Finally, to fulfil our commitment to evaluate every approach capable of producing body panels, we are considering other exciting materials and processes for this family of components. These include:

The thermoforming or sheet stamping of engineering thermoplastics

The stamping of plastic-metal laminates

Co-injection or sandwich molding, and,

The development of high-performance reinforced thermoplastic sheets.

Criteria for material/process selection

Design flexibility

Appearance

Performance

Productivity

Each exterior body and closure panel plays a role in establishing the appearance of the vehicle. Therefore, the selection of plastic materials and molding processes for these components must provide the capability to produce a variety of contours. As the range of vehicle size is reduced due to downsizing, the requirement for specialty models will definitely increase. The versatility of plastics and their effectiveness in producing specific design characteristics must be utilized to give designers greater flexibility in creating product differentiation features.

The appearance standards for exterior panels are well known, and plastic components must be produced with a surface quality comparable to conventional sheet metal. In addition to producing a smooth surface, there are obvious facilities and colormatch advantages in painting plastic panels „on-line" with the car body. The most desirable process is to install the plastic parts on the body prior to any finishing operation. At Ford, this means that materials must withstand several exposures to elevated temperatures, including the Electrocoat primer oven maintained at 205 °C. Front end panels or fascias made with urethane materials are painted separately with an elastomeric paint system, but our development of reinforced urethane fenders is aimed at painting these components with the conventional body enamel which is cured at 140 °C.

The implementation of plastics in exterior body and closure panels will not be achieved at the expense of product performance. Stringent material and product performance specifications must be met to satisfy the requirements of each application. Quality, reliability and durability are most important considerations in the selection of materials and product designs for these components.

A major factor in the concurrent development of several materials and processing techniques for exterior panels is the need to improve productivity. The importance of this need cannot be underestimated if we are to be successful in expanding the use of plastics. In this category of components especially, where we are competing with the high-productivity sheet metal stamping, we must develop the means to reduce the fixed and variable costs associated with current plastic product manufacturing.

Although this Symposium is devoted to the subject of plastics on the exterior of cars, components in the other categories referred to earlier should be mentioned briefly.

The substitution of plastics for steel in seating systems offers major opportunities for vehicle weight reduction. Blow-molded polyethylene seat components were mentioned earlier, and stamped glass-reinforced polypropylene bucket seat frames have been introduced in the United States. Polyester compounds are also candidates for some seat designs.

Rapidly changing developments in the area of high-strength composites provide potential for the application of plastics to structural and functional components. These developments include:

polyester and vinylester compounds containing 50 to 60 percent chopped glass fibers for components such as, bumper beams, radiator supports and road wheels, filament winding of resin-impregnated glass strands to produce parts like leaf springs with continuous, oriented reinforcing filaments, and,

pultrusion of continuous reinforcing filaments in a resin matrix for constantsection products such as, beams.

Major accomplishments have already been realized with plastics on the exterior of cars, but we are now entering new areas of application which require aggressive programs for the development of improved materials and processing techniques. The automotive industry, worldwide, must be committed to the conservation of energy, and the efficient use of materials, such as plastics, must be a primary objective.

The FOA Blackwood Falcon bumper system

Klaus C. Glagow

Introduction

In late 1973 FOA commenced the research, design and development of the 1979 Blackwood Falcon carline. The pressure to improve fuel economy through aerodynamic improvements and thorough application of lightweight design philosphy in combination with an existing strong plastics

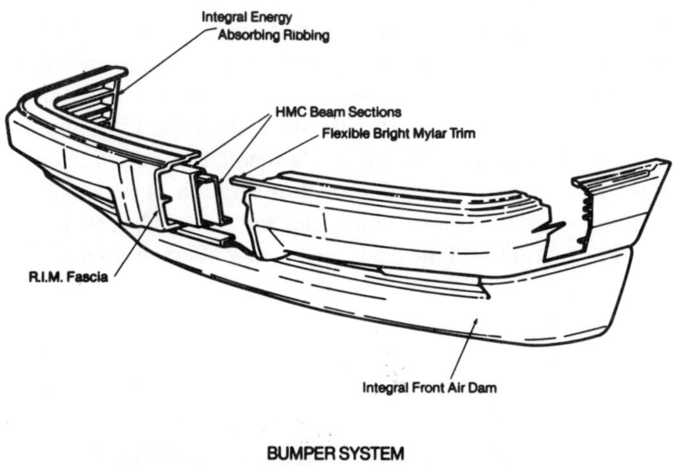

Fig. 1. Blackwood bumper system.

manufacturing division lead to the development of a lightweight energy absorbing bumper system. The system consists of a RIM polyurethane bumper cover with internal ribbing to achieve energy absorption, a lightweight HMC polyester fibreglass box section beam, bonded internally and to the PUR cover through use of PU adhesive, and HSLA Steel mounting brackets, fig. 1.

Product Considerations

An existing FOA in-house development standard required the bumper impact performance to meet a straight 4 kph barrier impact without any damage to adjacent body parts. In order to protect the body structure from the excessive input forces of a 4 kph impact, an energy absorbing bumper system was chosen. Extensive materials, process and assembly methods research ultimately left two possible choices for the principle to be used.

Davidsorb's System

This energy absorbing bumper system consists basically of a thin, decorative bumper skin moulded from PUR backed up by hard PU or PVC blocks supporting the skin cover and a load distribution beam to transfer impact forces into the body structure.

Bayflex System

This system features energy absorption through internally moulded ribs, supporting the heavy skin of the bumper, backed up by a load distribution beam. Since FOA had a 9000 m^2 plastics plant in operation since 1973, and plans for a 100 % expansion concurrent with Job 1 Blackwood, 'In-house' manufacturing of the bumper system became a mandatory requirement. This requirement was supported by additional integration of seat pads, steering wheels, crash pads, instrument binnacles and consoles; a less complex patent/royalty situation; and the ready availability of materials required in Australia, fig. 2.

Fig. 2. Plastics plant.

Design Considerations

The Bayflex system of energy absorbing bumpers had several applications in previous years. However, weight and design constraints demanded a fresh approach to this design principle.

The complete carline (Falcon, Fairlane, LTD, Wagon, Utility and Van) features 3 different front and 3 different rear bumpers. The Falcon Sedan, Wagon, Utility and Van use a common front bumper, with a unique bumper each for Fairlane and LTD. The rear bumper is common for Falcon Sedan, Fairlane and LTD with a unique bumper each for Wagon and Utility/Van.

All front bumpers feature integrally moulded lower air intake grilles, body colour painted integral lower valance panel and a flexible bright mylar strip inserted for decoration. The lower valance is not retained along the bottom since the PUR skin provides sufficient stiffness for self-support.

The Bumper section provides with its overhang a 60 mm protection zone for critical damage-prone areas like headlamps, grilles, turn indicator lamps and taillamps with a collapse zone of 45 mm to provide the energy absorption feature, figs. 3 to 8.

Fig. 3. Falcon front.

The appearance of the bumper makes the system an integral part of the body styling with sight shields overhanging from the rear edge of the bumper to eliminate see-through conditions. The front valance panel forms an air

Fig. 4. Fairlane front.

Fig. 5. LTD front.

Fig. 6. Falcon rear.

Fig. 7. Utility rear.

Fig. 8. Front bumper and valance.

dam to improve the aerodynamic character of the front end of the vehicle. The bumper wraps around the front fender back to the wheel opening to give good corner impact properties. A box section HMC (65 % glassfibre polyester bonded) beam consisting of a C and a Z section bonded together, using Goodyear Pliogrip 6000 PU adhesive systems is bonded by the same adhesive material into the heavily ribbed section of the bumper cover. Mounting arms of HSLA steel to minimise weight, attach the bumper assembly rigidly to the body structure.

Due to the usage of a common beam and mounting arms for all three front bumpers, with additional plan view curvature of the Fairlane and LTD bumpers, a PU filler in honeycomb configuration is inserted during assembly between beam and cover. This insert further improves impact properties for the higher weight car derivatives.

The Falcon sedan bumper system achieved a total weight reduction of 19 kg compared with a conventional chrome plated steel bumper system, with a concurrent improvement in bumper impact performance and overall body protection properties, figs. 9 and 10.

Fig. 9. Box section HMC beam.

Fig. 10. Beam bonded into front bumper.

Development, Prototypes and Design Verification

Due to design capacity limitations in Australia the detailed design program for the bumper system took place in the USA whilst the styling of the bumpers was retained in Australia. Prototype parts for the beam sections were produced from Kirksite dies by Budd in their Milford Plant in the USA. The bumper covers for both front and rear Falcon prototype bumpers were manufactured by Bayer in Leverkusen-Germany in their technological Laboratory. Initial bonding tests and sled impact tests were conducted in Germany with follow-up sled and pendulum tests done in Detroit, USA. Further bonding tests performed by Goodyear in their research laboratory in USA improved the quality of adhesive joints to a satisfactory level and lead to the final selection of the adhesive material and process.

However, performance and damage patterns on impacted bumper bars demanded a second generation of prototype tools. For the beam sections the production dies were advanced, and used prior to hardening to produce representative second generation prototype parts. The bumper covers were retooled again using epoxy prototype tools to produce updated parts to develop and improve the moulding process and to assist in the design of final production tools. Since fully representative prototype cars were available by that time, a full barrier impact program was conducted to validate the design in concept and details. Heat sag tests, vibration tests, UV exposure tests, corner impact, towing and wrecker towing, fit and function, general highway and accelerated durability tests, stone pecking tests on the painted portion, low and high temperature impact tests, rail and convoy shipment tests and flutter tests were conducted in accordance with approved Ford Design Verificated Procedures. The Design was 'debugged' using Failure Model Effect and Analysis methods.

A considerable number of changes had to flow back from the test and evaluation program to be considered and reflected in drawings, tool design and tool build. Some of the more important revisions related to provision of more generous radii of bumper arms facing the HMC beam to eliminate shear failure of the HMC under impact, and revisions to the fixing pattern and thickness of beam components for the same rear Stand-offs were provided on HMC beam components and covers to give precise adhesive gaps. Additional webs in diagonal configuration had to be added to the wraparound portion of the rear bumper to eliminate heat sag under paint repair oven temperatures (120 °C for 20 mins).

Various revisions to the webbing pattern on the inside of both front and rear bumper covers took place improve load distribution and reduce shear forces in the adhesive joints under impact, and to improve material flow and ventilation of the mould during the injection process. Further revisions related to the insertion and clipping arrangement of the decorative mylar strip and end fixing of this strip. Considerable effort was required for the development of double sided adhesive tape to assist the clipping feature in retention of the mylar strip.

Since all major components of the bumper system were using new design concepts, new manufacturing techniques and equipment and new assembly methods, a high degree of caution was displayed by all activities involved. Thanks to the excellent co-operation of Goodyear US, Expandite Australia, Budd US, Escon Australia, Mobey US, Bayer Germany and various activities of the International Ford Organisation like Plastics Development Centre, USA, Engineering and Research Staff, and Overseas Product Engineers, the Engineering and Development program was completed on time with satisfactory results.

Open issues remaining at this time were UV stability of unpainted material, serviceability of a fully bonded bumper assembly, and customer acceptance.

Manufacturing Process

Bumper Covers

Since the Bayflex 91 material had been chosen for both front and rear covers the specified process data for this material had to be adhered to. However, due to early UV exposure sample failures chemical modifications to the Polyol were formulated by Bayer, requiring minor process adjustments, fig. 11.

Mould Preparation

A spray coat of mould release agent is applied on the upper and lower portion of the die. After extensive and costly experiments with approximately 15 different types of material, a wax base AKMOS 180 ST5 was selected giving acceptable results in mould release and part cleaning. However, it must be emphasised that the selection of the mould release agent constitutes probably the most critical process factor in the PUR bumper cover manufacturing process, and the end result, quality of components and repair and scrap rate can depend completely on the selection of this release agent. The experience gained in Ford of Australia is strictly

Fig. 11. Bumper cover as moulded with sprue and flash.

related to galvano-formed nickel/copper shell type tools and may vary widely with different tools being used.

PUR Injection

Injection points have been chosen at the top inboard edge of the covers at centreline of car. The Bayflex standard mould mixing configuration with a 'U' shaped material path is used. Tool gaps are maintained at minimum to reduce mould flash. Injection time is between 1.85 – 4.1 seconds depending on the cover moulded a 90 second cure time. The initial tool tryout revealed mould finish and ventilation problems resulting in unacceptable reject rates. Time and effort was spent to rectify the identified problems and production is now running steady at 3 % reject and 7 % rework rate for minor surface deficiencies with small surface voids constituting the major cause for the rework. Further improvements in mould ventilation and design improvements are expected to reduce rework rate to around 3 % and reject to less than 1 %.

Demoulding and Part Trimming

Parts are demoulded manually by one operator, and cooled off for approximately 120 seconds whilst the operator prepares the mould for the next shoot. Flash and sprue are removed by hand using special knives.

Production Volume

Since one front and one rear bumper cover are moulded concurrently in one press and one shot, a cycle time of 14 parts each per hour is continuously achieved. When only one tool is used in the press 18 parts per hour are produced.

A minimum of 45 minutes and up to 60 minutes per shift is used to clean the mould thoroughly and remove wax residue using a blended solvent-based cleaning agent Chem-Trend 101. Thus, only 7 hours per shift production time can be maintained, giving a maximum of 378 pairs of covers per production day.

Part Preparation

To prepare the PUR covers for painting and adhesive bonding, a 4 minute vapour dewaxing process using Trichlorethylene is used. Cleaning is critical and determines the adhesion qualities of both paint and PU adhesive. Lengthy experimentation and variation of nozzle positions and pressure was required to achieve optimum cleaning results on inside surfaces. Since front covers absorb up to 180 grams of Trichlorethylen during the cleaning process, good post-cure and subsequent ventilation of parts is essential. The post-cure is over 30 minutes at 115 °C on formed jigs to stabilise parts and support large surface areas for minimum warp and controlled shrinkage. Post-cure will also drive residual blowing agents out. However, small amounts of both blowing agent and Trichlorethylene still remain in the PUR cover, and only disappear after painting.

Painting

PPG 100 primer is used after masking and minor tack-off solvent wipe. The primer is cured at 110 °C for 30 minutes followed by 3 top coats (apply wet on wet) of PPG 100 cured for 30 minutes at 110 °C. Paint touch up is required on approximately 5 % of paint front bumpers, with surface voids showing up after repeat heat treatment. Repair is conducted by filling of the voids with elastomeric material and paint touch-up with PPG 100, followed by further curing cycle of 30 minutes at 110 °C.

HMC Beams, fig. 12

Both the C and Z sections are moulded from high strength moulding compound consisting of approximately 65 % chopped glass reinforcement and approximately 33 % polyester, with minor content additives to improve

Fig. 12. HMC beam components.

mouldability Placement of prepreg. material lumps and shoot weight is critical for orientation of glass and consistent thickness of the moulded part. Since considerable experience existed in-house with large size SMC front end panels and applied SMC rear panels, production launch problems were minimal. However, these components are the first worldwidc application of the HMC material developed by the Budd Compana and locally produced in Australia under license.

Four parts (2 front and 2 rear) are moulded concurrently in one press stroke with a 105 second curing time in the tool. No mould release agent is used since HMC material contains additives assisting mould lubrication. After demoulding, parts are hand trimmed in a minor trimming operation, and fixing holes, although premoulded, are pierced using a small hydraulic press. The parts are used without further cleaning for the bonding operation.

Assembly, fig. 13

Although the PU adhesive bonding process is extensively used in several applications (shoe industry, fibreglass cabins etc.) these applications are either on small parts or in low volume production. Since appearance consideration and concern on screw tension relaxation excluded any conventional method of assembly, the components of the fibreglass beams as well as the beam-to-cover joints are bonded using a Goodyear Pliogrip 6000 Polyurethane adhesive manufactured by their licensee in Australia. Since the adhesive joints and their integrity determine the ultimate impact performance for the overall bumper system, considerable time and effort was spent

Fig. 13. Assembly fixture roundabout.

to develop an adequate manufacturing technology to achieve consistent production quality. Cleaning of the bonding surfaces is most critical for adhesion performance and is achieved through a double solvent wipe after the covers have already been thoroughly cleaned during the paint process. Adhesion primer as specified by the adhesive manufacture is applied to all bonding surfaces. A continuous bead of premixed adhesive is then applied to one of the components.

The fibreglass beams are bonded at room temperature; the bumper covers are preheated to 90 °C to improve consistency of bonding. The components are now clamped and kept under clamping pressure for a minimum of 15 minutes. Clamping is then removed, but ultimate performance of the joints is not achieved for another 48 hours of maturing. The beams are assembled initially, and completely bonded beams are bonded into bumper covers. The decorative mylar strip is inserted into the covers prior to bonding, since at each end, staples or drive screws from the inside prevent end-lifting.

Mounting arms are attached by the Z section of the beam by means of J-nuts and bolts. The fully assembled bumpers are shipped in mobile racks to the assembly plants.

Tools and Facilities, fig. 14 to 16

Covers

The cover tools are galvanoshells (Nickel/copper electroforms) backed up by aluminium filled epoxy resin held in a one-piece aluminium cast support. Cooling tubes are brazed onto the galvanoshell. Cavities for webs and reinforcements on the back face of the cover were spark-eroded onto the galvanoshell. Galvanoshell of such a physical size and complexity are not commonly used and a considerable number of problems had to be resolved during tool design and tool build. The required coordination between Australia and Germany and the number of tool changes emerging from the test and evaluation program to be reflected in the tools made the timely completion of the dies a formidable task. Timing constraints demanded production start-up and some fine timing was required for the first 4 – 5 months of production to achieve the projected production results.

New Kannegiesser tilting platten type RIM presses are used. These presses take one front and one rear bumper cover tool each. Only minor problems were experienced with the hydraulic turn motors. However, this type of press requires two operators to apply mould release agent and for the mould cleaning operation.

Fig. 14. Galvanoshell.

Fig. 15. Die ready for epoxy casting.

Fig. 16. Die open.

Fig. 17. Press open with one front and one rear die in place.

Fig. 18. Polyol and Isocyanate pumps.

Hennecke mixing heads and HK 1000 pumps are used for Polyol and Isocyanate injection, fig. 17 and 18.

Beams

Steel dies are provided to mould the HMC beams. Although each part has its own individual die, four dies are mounted into the press to provide 2 front and 2 rear beam components concurrently. These dies were completed very early during the development program and left soft. Prototype parts for 2nd generation prototype bumpers came off these soft production dies. All resultant changes from the test and development program were included into the dies prior to hardening. A 600 tons hydraulic press is used to mould the HMC beams. Theoretically the press tonnage should be 25 % higher, but no major problems have been encountered.

Cleaning and Painting, fig. 19

A designated vapour cleaning facility had to be installed to achieve the required dewaxing and cleanliness for adequate paint and adhesive performance. A post cure oven with fume extractors to extract the cleaning medium vapours is provided.

Fig. 19. Spray booth exit.

A spray booth is provided for PPG 100 primer and top coat application with a choice of up to 14 colours. The primer and paint cure ovens, as well as the post-cure oven, feature indirect firing, since volatiles from RIM moulding process and paint process combined with an open flame cause development of hydrochloric acid, creating frosting effects on paint and PUR materials, fig. 20.

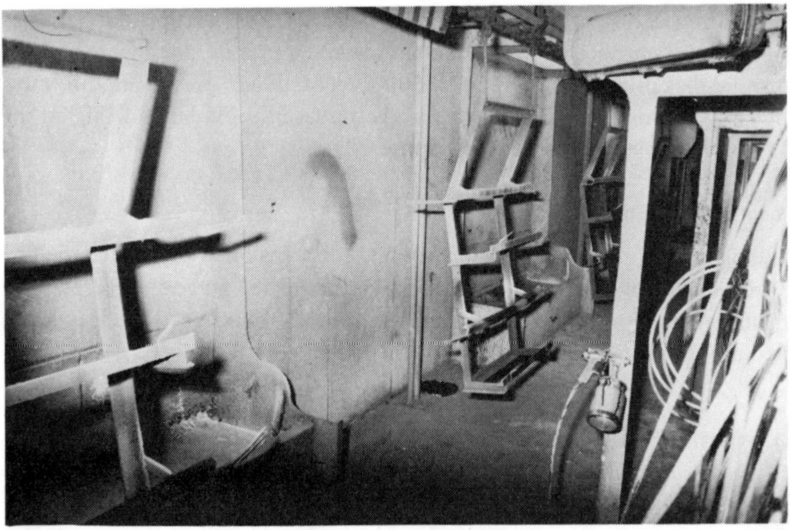

Fig. 20. Spray booth inside.

PUR Bonding

Hermex PUR adhesive low pressure dispensing equipment is in use, as recommended by Goodyear. The application onto components as manual through nozzles. Extensive operator training is essential to give good and consistent bonding results.

The dispensing equipment requires regular flushing with methylene chloride, considered a hazardous material. Adhesive primer application requires good ventilation due to toxity. A Roundabout assembly fixture set up is installed to achieve the required bonding pressure with a preheating station to improve bonding strength and curing results.

Production and Launch Problems

The adoption of a completely new design with the concurrent commissioning of an all-new production facility, with the additional difficulty generated through the necessity of worldwide communication and technical liaison, made the accomplishment of a timely job 1 and good production acceleration a very demanding objective. Numerous problems were resolved prior to production start up. However, production is now well underway and comes up to expected quantity levels. Some possible technological and tooling improvements are identified and would be realised whenever a new program offers the opportunity.

Some of the specific problems encountered related to the design and execution of the bumper cover moulds. Inaccuracy of tooling aids caused dimensional discrepancise along the groove section which houses the mylar strip, This problem generated retention problems for the decorative strip, with subsequent development work required to release double-sided adhesive tape to provide the major mechanism for the retention of this strip. Timing problems caused part of the mould, related to the rear face of the cover, not to be polished as specified. This in turn made demoulding very difficult, and residual PUR build-up in moulds created excessive downtime for cleaning operations. The spark erosion used to form the webbing pattern on the rear face of the bumper generated separation of the galvanoshell with subsequent fatigue fracture of some thin wall sections of the mould. A general deficiency of the galvanoshell is that the nickel surface can not be reworked or repaired without jeopardising the whole integrity of the mould. The RIM presses as procured proved to be to complex in their operation, through their tilt features. Simple clam type presses would improve cycle time and access to the tool. They would further remove the need for two operators for cleaning and wax spraying. Pumping equipment proved extremely prone to failures, and so some pumps have had to be replaced due to bouncing of pistons, seal problems and minimum clearances partucilarly on polyol pumps, causing blockages of these pumps. Mixing heads as installed did not fully meet the demanding requirements of 24 hour operation. Sticking needle valves and throttling valves, controlling recirculation, require continued adjustment. Even the firing pressure of neddle valves requires frequent adjustment. Later generation equipment should be capable of reducing downtime through less complex, more practical designs. The assembly operations suffers from the inadequacies of dispensing technology for PUR adhesives.

Production proof, low wastage dispensing equipment is presently unavailable. The equipment available either generates mixing ratio variation in cycling application, or in continuous operation generates approx. 45 % material loss with no opportunity for recovery. The dependency of bonding quality on the operators' quality of workmanship is very high, and control of the end result can only be verified through destructive testing.

Materials Problems

Bayflex 91 Components

Control and storage of the raw material components of Bayflex 91 is highly critical for successful operation. Experience with shelf-life indicate considerable variations. An extension of shelf-life would be desirable in view of the long lines of supply (Germany – Australia). Temperature control requires high accuracy. ± 2 °C at the mixing head should be the maximum tolerance to achieve the specified reaction and avoid mixing problems. Isocyanate is prone to low temperature crystalisation, and reduction of temperature below + 10 °C can generate this effect. Local production or local blending of materials is required to reduce problems as experienced. Any interruption of the present line of supply necessitates airfreight, and with daily material consumptions of up to 8 tons, this is not economical.

Aeration of Bayflex 91 is required to cause correct foaming. Insufficient aeration of material in their pressurised containers will generate incorrect material reactions.

Users in the USA and Europe are using devices to measure and dose aeration.

Mould Release Agent

The presently used material is giving acceptable results. However, this material is highly critical relative to the total future of PUR for complex components, and improvements are definitely required for better production results. In-mould painting might open up new avenues for mould release agent technology.

Adhesion Primer

This primer is used on HMC and PUR to improve adhesion of PUR adhesive. The application process and cleanliness of surfaces are highly critical, and the toxicity of the material does not encourage experimentation.

No other materials used have shown any significant problems in production.

Cost Considerations

Compared to a conventional chrome plated steel bumper system, no major piece cost saving has been achieved. There is however, some scope to reduce cost through simplification of bonding processes or substitution of bonding. Further potential exists in reduction of HMC beam thicknesses or simplification of these beams. $ A 1.4 million has been saved in the tooling program of the new bumper system, despite tooling cost increases through design and manufacturing requirements; an important factor for relatively low volume production.

Field Experience

Public and press response at and after public introduction of the Blackwood model range, related to the new bumper system, has been excellent. After a full year of field experience the status can be summarised as follows:

Impact Behaviour

Minor impacts are easily absorbed by the system, leaving in the worst cases some minor surface blemishes. Major impacts of a magnitude requiring full replacement of a conventional bumper system and extensive body repair work are usually restricted to partial repair of the HMC beams, and rebonding of the covers. Comprehensive service repair procedures have been released to Dealers, with a range of repair materials readily available. Even in impacts at 50 km/h as conducted for steering colum intrusion compliance, and rear end impacts of similar magnitude, the bumpers does not detach from the vehicle, although damage experienced under these conditions is certainly beyond repair capability.

Serviceability

As indicated above, minor damage resulting from impact or surface blemishes can be repaired in service. However, due to the energy absorption feature the bumper system is less prone to suffer any damage in minor impacts. Since the complete bumper is easily detached from the car through its 6 fixing screws, and beams and adhesive joints are then readily accessible from the back, inspection and repair work pose no problems.

Ultraviolet Light Degradation

Painted Surfaces

Since well developed paint systems are used, the UV degradation of the painted surfaces of the front valance was never considered to pose a problem. Although the PPG 100 paint system had not been used on Australian motor cars before, fading or loss of gloss level after one year of exposure was minimal, and showed results equivalent to orbetter than conventional body enamel systems. Since no research data for comparison of UV load between Florida/ USA and Allunga/Queensland Australia is available, we consider at least a 40 % increased load a realistic figure.

Unpainted Surfaces, fig. 21 to 23

Neither the horizontal surfaces on top of the bumpers nor the vertical impact surfaces, are painted. Concern existed after 3 and 6 month exposure results of standard Bayflex 91 material samples in Allunga. Modified polyol formed the basis for the next series of tests, with no more encouraging results. The surface finish of the critical surfaces was modified to satin vapour honed to alleviate the effects of loss of gloss level. Samples with 1 % Tinuvin used as UV inhibitor, did not show any significant improvements. Regular waxing of exposed surfaces with various commercially available wax materials and waxes containing UV inhibitors was equally ineffectives. Extensive studies on competitiv cars in Australia and the South West of USA using

Fig. 21. Modified polyol 4 month exposure.

Fig. 22. Standard Bayflex 91; 7 month exposure.

Fig. 23. Bauflex 91 with 1 % Tinuvin stabilised; 7 month exposure.

PUR exterior components with up to 5 years of service life, lead to the final decision to discontinue further attempts to improve UV stability. A recent field survey in Northern and Central Queensland (Australia's highest sunload area) of 11-month old Blackwood cars confirmed the initial decision, that UV degradation on cars in service is far less obvious than would be indicated by test samples in exposure tests, and that normal owner care (washing and waxing) with standard car care products will alleviate the effects of UV degradation on unpainted Bayflex 91 bumpers and maintain acceptable appearance over the normal expected service life of the car.

Outlook

The obvious success of the energy absorbing Blackwood PUR bumper system will encourage further application of such systems.

We are proud that Ford of Australia was awarded an Honourable Mention award for last year's most innovative use of plastic by the Society of Plastic Engineers (Plastics in Automotive Division) in Detroit in October 1979.

Future generations of bumpers will show various design improvements. Newly developed high-modulus and glass-filled PUR materials will enable the engineer to simplify the design and reduce cost and weight further. Manufacturing technology will require further incentive to give improved production results, and considerable advancement can be foreseen in refining today's equipment. The next 5 years will show how the challenge of producing large PUR parts at low cost and in high volume can be met.

A flexible SMC for automotive applications

Kenneth A. Iseler and James N. Quinn

History is often important in understanding the present. In that light some background on the technical development of a tougher, more flexible SMC will be presented. Following this presentation, some of the trends seen in the United States for automotive SMC applications will be reviewed.

Now back to the history behind the development of the tougher, more flexible SMC. Five years ago the Budd Company began work to develop a more impact resistant SMC having more uniform glass distribution and as a result more consistant physical properties. If SMC were to continue to grow as a material of design choice by automotive engineers as a replacement for metal, its toughness and uniformity had to be improved. The first step was successful and three to four years ago the resulting material became the standard low profile SMC moulding compound of the Budd Company.

At that time the major automotive use for SMC in the U.S. was in grille opening or front end panels. However, in order to successfully open up the market for other automotive exterior body panels in a big way still further improvements in SMC would be required. At the same time the oil embargo in 1974 intensified the demand in the U.S. for weight reduction and further put the spotlight on lightweight substitutes for steel such as plastics and aluminium.

The timing appeared to be excellent for SMC for large body panels, and even more attractive if a tougher, more impact resistant material could be developed to permit the production of thinner, tougher parts; but without sacrificing the excellent dimensional stability, class "A" surface, and paintability of SMC in the auto manufacturers' paint ovens.

About the same time other plastic materials such as RIM (Reaction Injection Molding) also began to be recognized as a possible material for the replacement of sheet metal in exterior body panels. The producers of RIM parts began active programs to try to introduce glass reinforcement into their materials to make them stiffer and more dimensionally stable. Aluminium producers increased their efforts to develop more uniformly stampable, dent resistant alloys, and The Budd Company as a SMC producer began the second phase of the development to produce a still tougher but more flexible SMC. The race was on!

Sheet molding compound

Sheet molding compound (SMC), as known in the plastics industry, refers specifically to chemically thickened unsaturated polyester resin containing inorganic filler and chopped glass reinforcement. Thermoplastic polymers are added to the polyester when low shrink, low profile surface characteristics are required. The SMC formulation also contain initiators to promote polymerization and internal release agents. In the manufacture of SMC, resins, filler, initiators, release agents, chemical thickening agents, pigments, etc. are preblended and dispensed onto the SMC machine (fig. 1 shows a schematic of the SMC machine). Glass roving is chopped to length, typically 2.54 cm., and "worked" by various means to impregnate or to wet out the glass bundles. Glass percentage can be varied according to the mechanical properties required. The SMC is then maturated at controlled temperature and time to allow for proper viscosity build to reach good handleability and glass-carrying ability. Typically, the SMC is moldable after a few days and remains moldable up to a few months. The SMC is molded in matched metal molds at a typical temperature of 150 °C and pressures of 70 kg/cm². A molded part with a wall thickness section of 3.175 mm will typically cure in 60 seconds.

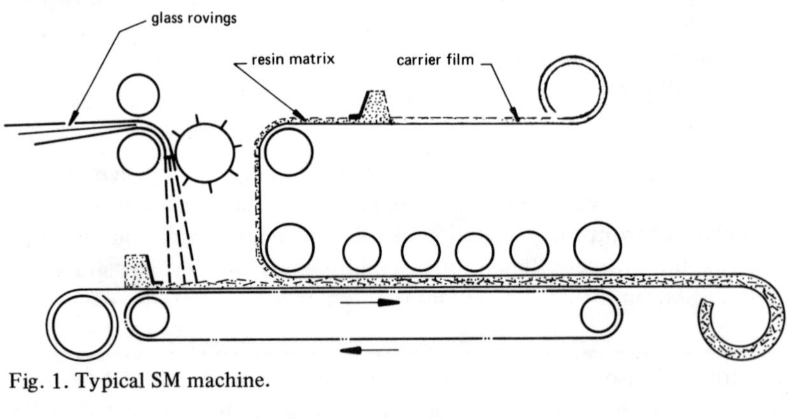

Fig. 1. Typical SM machine.

SMC moldings, while satisfying the strength requirements necessary for body panels, has had a major drawback in the lack of flexibility. The need to conform to the variation in metal stampings, to which in many cases it is attached, has often times resulted in cracking. This is especially true when the necessary racking and twisting occurs during mounting to the sheet metal. Needless to say the SMC industry has been looking for a flexible SMC for many years.

Objective

Our objective in the development of Flex 2000 was to develop a sheet molding compound that was 10 – 15 % more flexible than conventional SMC without sacrificing any of the fine characteristics that have attributed to the phenomenal growth of SMC such as:

1. Dimensional stability.
2. Low shrink-low profile appearance.
3. High heat paint requirements i. e. 190 °C.
4. Good physical properties.
5. Processability and handling.
6. Spiral flow for mold fillout.
7. Glass-carrying ability.

Experiments

Many experiments were designed to evaluate various additive, modifiers, monomers and polymers to improve on toughness or flexibility. Experimental formulations were developed and the SMC matrix was evaluated for viscosity build. Laboratory panels were molded and evaluated for cure, surface appearance and shrinkage. Only the promising formulations were made into SMC "mini-runs" on our laboratory SMC machine. These "mini-runs", which contained glass reinforcement, were then further evaluated for various physical properties.

Our initial test, to screen many variables, was the modulus of elasticity. This was performed on laboratory panels as molded and again after subjecting to automotive paint cycles of 190 °C for 30 minutes for 3.175 mm thick moldings.

It was found that some formulations gave us a lower modulus of elasticity, but fell short in other necessary criteria as previously outlined. The test, where the majority of promising formulations failed, was the inability to withstand high heat. Some systems post-baked i. e. underwent further polymerization during the high heat. Other systems showed signs of polymer degradation upon exposure to 177 °C. As development progressed, new polyester chemistry began to yield results in the direction of achieving our objective. The new unsaturated polyester resin, the composition of which is proprietary, is the basis of Flex 2000.

Flexibility increased 25 %

With refinements and modifications in formulation technology, where we took into account the broad scope of SMC, we were eventually able to come up with a molding compound that, when molded, was 25 % more flexible than conventional SMC. This far exceeded our original objectives of a 10 – 15 % improvement. The increased flexibility was measured by a reduction modulus of elasticity from a typical value of 11,730 MPa to 8,970 MPa. Physical property analysis, other than tensile and flexural modulus, showed little or no change. Fig. 2 compares the typical properties of a conventional SMC with Flex 2000. Where some of the properties such as specific gravity, water absorption, heat deflection temperature, mold shrinkage, coefficient of expansion, notched izod impact, flexural strength, and tensile strength remain unchanged, Flex 2000 however, shows marked improvement in the modulus of elasticity (flexural and tensile modulus).

Properties	Conventional SMC	Flex 2000
Glass Content	28 %	28 %
Specific Gravity gm/cc	1.9	1.9
Water Absorption (24 Hr. at 23 °C)	< 0.35 %	< 0.35 %
Heat Deflection temp., °C, 1.82 MPa	> 177	> 177
Barcol Hardness	55 – 60	45 – 50
Mold Shrinkage, cm/cm	0.000 – 0.003	0.000 – 0.003
Coefficient of expansion, cm/cm/°C	19.8×10^{-6}	19.8×10^{-6}
Compressive Strength, MPa	166	152
Izod Impact Strength, kJ/m, notched	0.96 – 1.17	0.96 – 1.17
Flexural Strength, MPa	179.4	179.4
Flexural Modulus, MPa	11,730	8,970
Tensile Strength, MPa	72.5	72.5
Tensile Modulus, MPa	12,420	9,315

Fig. 2. Flex 2000 comparison with Standard SMC.

New test instruments

However, two new test instruments were very beneficial in the development of Flex 2000. Data from the Dynatup Impact Tester and Acoustic Emission Instrument was extremely beneficial in the fine tuning and properties characterization of Flex 2000. In the future, as new compounds are developed, these two instruments will play an important part in performance differences.

A brief summary of these two test methods are described as follows:

Dynatup: The word dynatup is a registered trademark of Effects Technology Inc[1]). It is derived from "dyna" which means dynamic and "tup" which means impacter or striker. This is a drop weight system which consists of two parts, an impact machine and an automated instrumental impact system. Dynatup Model 8200, an impact machine in conjunction with Dynatup ETI-300 (see fig. 3) is an automated instrumental impact system used to verify "toughness" for the Flex 2000 as compared to the conventional SMC

1) Effects Technology Company is represented by Roell and Korthaus DG Düsseldorf, West Germany.

Fig. 3. Dynatyp ETI-300.

under similar conditions. All the test specimens were tested at the same thickness at 3 mm, since thickness plays a very important roll in the results obtained. This Dynatup system not only monitors the entire impact event from initial impact through fracture, but prints out precise load vs. deflection (or time) and, energy vs. deflection (or time) curves based on the data.

A typical curve of load vs. deflection and energy vs. deflection based on the data for conventional SMC and Flex 2000 are shown in fig. 4 and 5 respectively. All test data is printed out automatically by the automated instrumental impact system. Both the conventional SMC and Flex 2000 were tested at impact velocity of 2.82 meters per second. Fig. 6 is a schematic of energy vs. deflection comparing Flex 2000 to conventional SMC. Flex 2000 is 55 % more energy absorbing than conventional SMC. Fig. 7 is a schematic of load vs. deflection comparing Flex 2000 to conventional SMC. Maximum

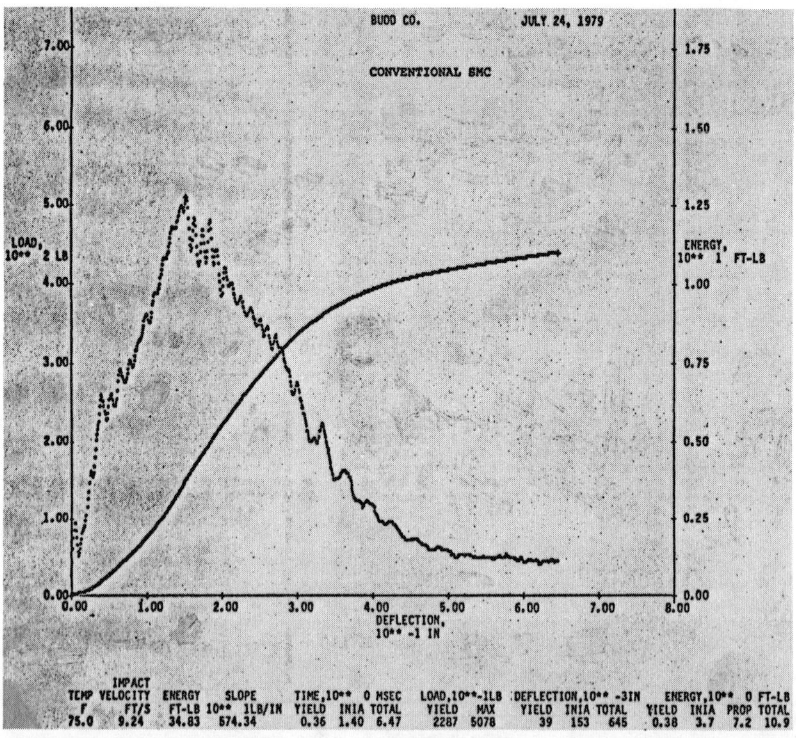

Fig. 4. Typical curve of load vs. deflection and energy vc. deflection based on the data for conventional SMC.

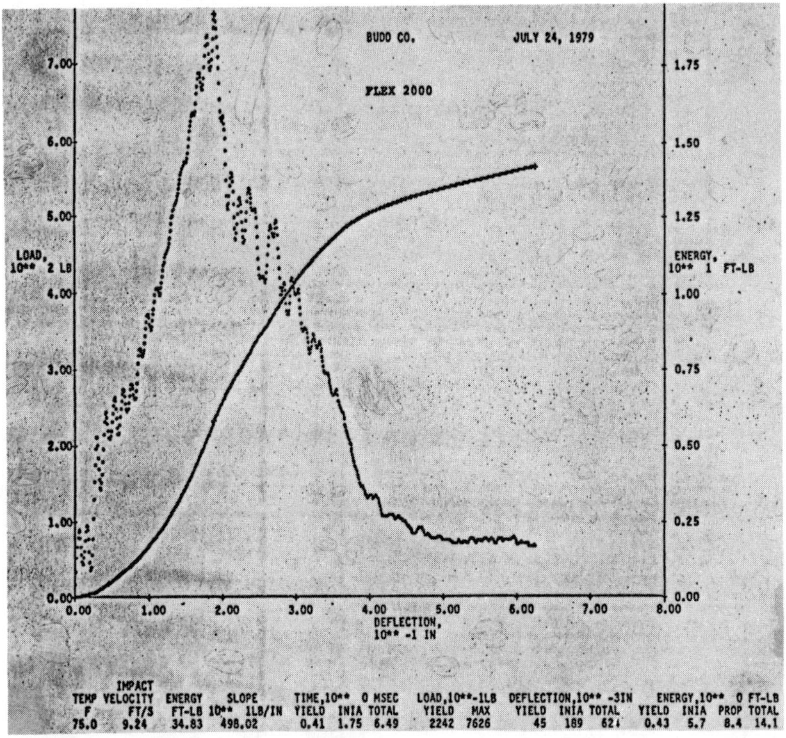

Fig. 5. Typical curve of load vs. deflection and energy vc. deflection based on the data for Flex 2000.

force for Flex 2000 is 50 % greater at fracture than conventional SMC. Although there is much data available from these previous curves, fig. 8 summarizes the most important factors. From this Table, it is seen that Flex 2000 shows about 55 % higher energy absorption, i. e. tougher; and takes 50 % greater force to fracture as compared to the conventional SMC. Also, Flex 2000 takes about 25 % longer to initiate fracture and, because of its flexibility, deflects about 23 % more at maximum load over conventional SMC.

Acoustic Emissions: Acoustic emission counts can be defined as the actual number of sound waves generated from a test specimen under stress. Acoustic emission counts vs. load is plotted in fig. 9 for the conventional SMC and Flex 2000. It is shown in the curve that Flex 2000 can exhibit stress up

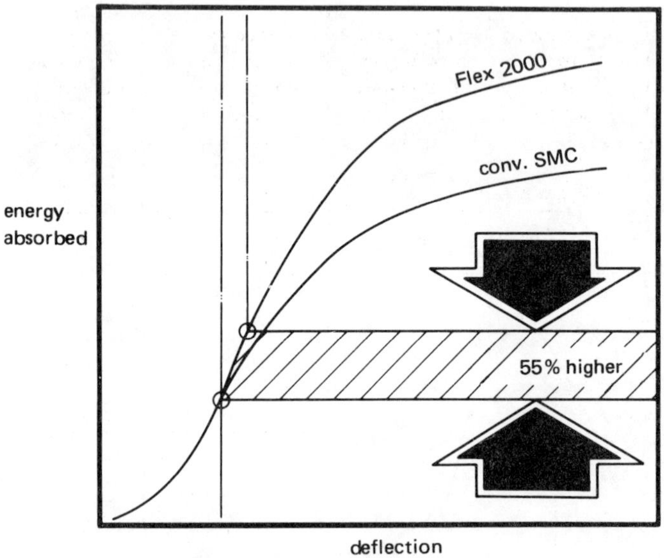

Fig. 6. Energy absorbed (at maximum deflection).

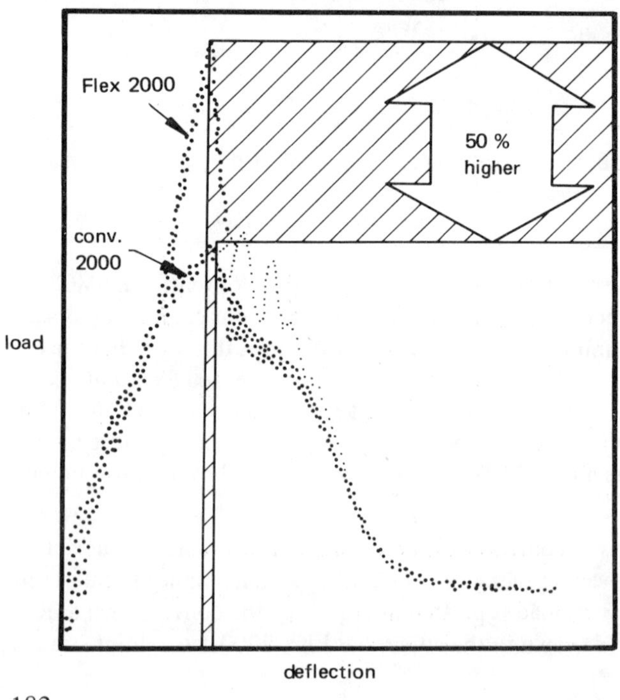

Fig. 7.
Maximum Load
(at yield point).

Description	Time required to initiate fracture	Maximum force obtained by specimen at fracture	Deflection occurred at maximum load	Energy absorbed by specimen at initial fracture
Conventional SMC	1.4 millisecond	230 kg	3.9 mm	5.0 J
Flex 2000	1.75 millisecond	346 kg	4.8 mm	7.73 J
Flex 2000 improvement over conventional SMC	25 %	50 %	23 %	55 %

Fig. 8. Comparison of impact and deflection.

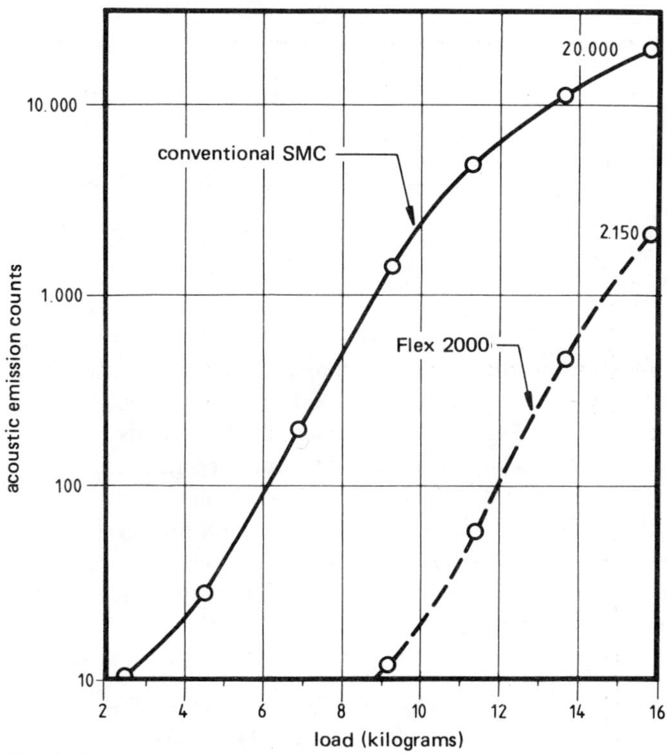

Fig. 9. Acoustic emissions-conventional SMC vs. Flex 2000.

to 8.16 kg without generating any acoustic emission counts whereas conventional SMC generates 750 acoustic counts. Acoustic counts reported, on the curve, are the total emissions generated from resin matrix crack, fiber debonding from resin matrix and actual fiber crack as it is beyond the limit of this test method, at present, to differentiate individual emission counts for the specific types of crack. Acoustic counts for Flex 2000 at 11.34 kg load are only 58 and at 15.88 kg load are 2,150 as compared to the conventional SMC which shows acustic counts of 4,840 at 11.34 kg load and 20,000 at 15.88 kg load, fig. 10. Thus, the difference is quite significant. Acoustic emission deflection is the deflection that occurred, in the test specimen, when subjected to stress. It can be seen that Flex 2000 deflects up to 22° as compared to conventional SMC that deflects only 13° at the same load of 15.88 kg.

Properties	Conventional SMC	Flex 2000
Acoustic emission counts		
at 6.8 kg.	204	0
at 11.34 kg.	4,840	58
at 15.88 kg.	20,000	2,150
Acoustic emission deflection	13°	22°

Fig. 10. Acoustic emission vs. load.

Scale-up

From laboratory pilot quantities to scale-up on production manufacturing SMC equipment, Flex 2000 processed and performed very well. Molding results were excellent when grille opening panels, fenders, hoods and quarter panels were molded on production presses. An important advantage noted during production trials was that Flex 2000 carried the glass fibers better than conventional SMC, as demonstrated by more uniform physical properties and better knitting characteristics. Since Flex 2000 is much more flexible and energy absorbing than conventional SMC, a program was initiated to reduce molded part weight by reducing part thickness.

Experimental moldings indicated that parts molded very satisfactorily when the wall thickness was reduced from 3.2 mm to 2.0 mm at no sacrifice in surface smoothness. This means a possible weight savings by as much as 33 %.

Summary

Improved properties now help to open up the market to SMC for certain high stress vehicle exterior body panels. Typical of these in truck applications would be fender and hood combinations and grille opening panels. In automotive, typical applications would be hoods, deck lids, doors, and fenders. Part and vehicle weight can be reduced through the use of tougher, more flexible SMC such as Flex 2000 because thinner, lighter wall sections can be molded without loss in structural integrity.

Some of the trends seen in the U. S. for automotive SMC will now be reviewed.

Trends

The market growth for SMC has been a rapid one in the U. S., as shown in fig. 11, averaging over 31 % per year in the U. S. over the last five years.

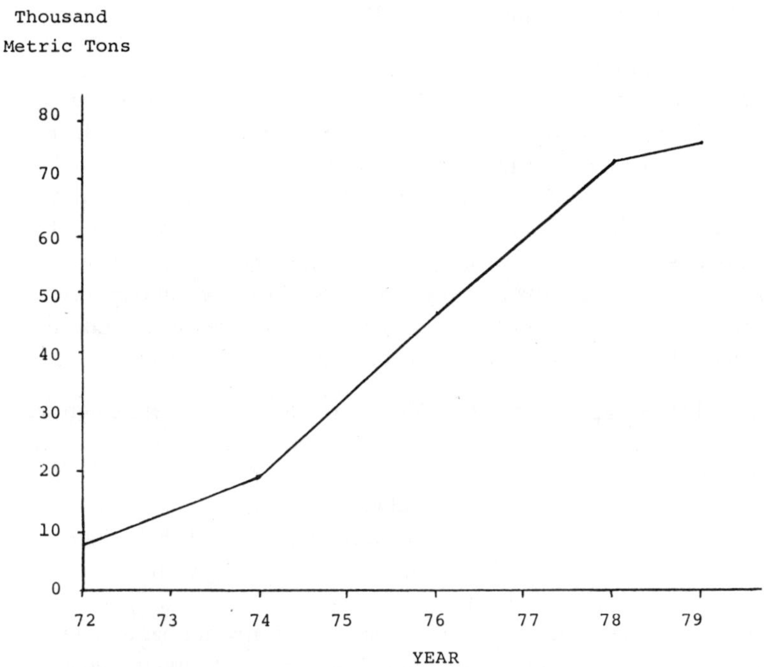

Fig. 11. SMC growth history U. S. automotive & truck only.

Car model	Capacity planning volume parts/year
Ford Ltd GOP	475,000
Ford Thunderbird GOP	330,000
Chrysler LeBaron	177,000
Chrysler Cordoba	70,000
G.M. Chevrolet Caprice	509,000
G.M. Oldsmobile Cutlass	530,000
G.M. Pontiac Grand Prix	260,000
Others	3,926,000
Total	6,277,000

Fig. 12. U. S. 1979 Model year SMC front ends.

As dicussed earlier, the big growth in that market was in front end panels to the point that in 1979, 6.3 million, or 65 % of all U. S. cars including almost all of the largest volume models have SMC front ends, fig. 12.

Now, however, with the increased emphasis on weight reduction coupled with the availability of improved SMCs, the use of SMC in automotive and truck applications are forecasted to grow at an increased rate over the next five years as shown in fig. 13. You can see an average projected growth rate for trucks and cars of 24 % per year in the U. S. This compares to 16 to 18 % per year in Europe. There is a higher growth rate for SMC in the U. S. vs. Europe because of the greater governmental pressures there for weight reduction coupled with the possibly greater weight reduction opportunities. However, the weight reduction thrust appears to be increasing in Europe also which should further help to spur the growth of SMC here.

The next big new applications for SMC to produce this accelerated growth rate are large truck and automotive exterior body panels. In the U. S., many truck exterior parts are already in SMC such as the front end and hood of General Motors "General" and "Brigadier" heavy duty truck models. The Ford "Louisville" and "W" series trucks also have most of their "below the beltline" parts molded from SMC. And many of the heavy duty truck manufacturers are actively considering redesigning into SMC not only for hoods and front ends, but entire cab, and International Harvester has announced their intention to build an SMC molding plant to produce their own SMC truck parts.

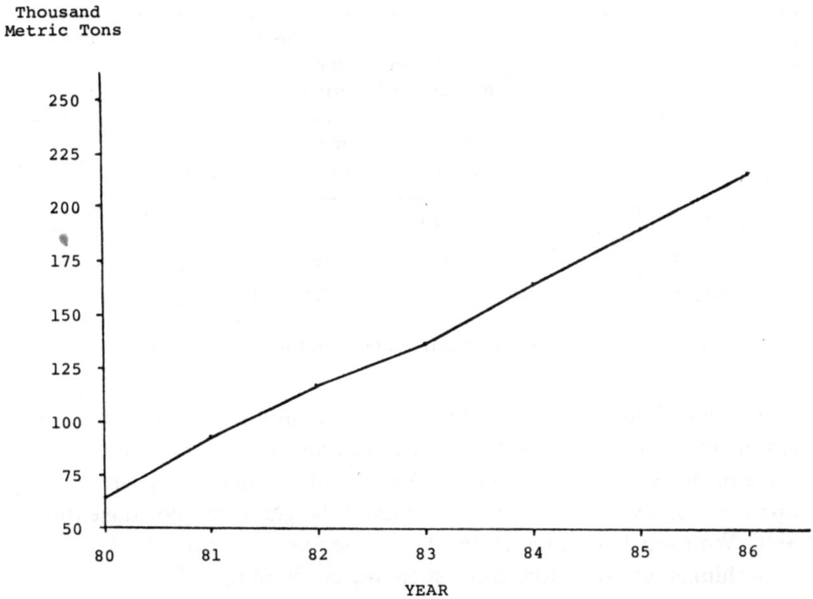

Fig. 13. Projected SMC growth U. S. automotive and truck only.

Regarding new automotive applications in the U. S., molds are being built now for production volumes of doors, hoods, roofs, and lift gates (or rear "5th" doors).

It is this enormous exterior body panel potential SMC market that has led to our committment to produce tougher and more flexible SMC. It is needed to help SMC continue its long term growth as one of the best plastic materials for exterior body panels. Basically SMC is an excellent material of design, as outlined in the beginning. In addition it is relatively low in cost, is versatile in that it can be formulated and reinforced to meet specific application requirements, and is energy efficient. SMC contains only 30 to 35 percent petroleum based products, although that is not really the important factor in determining energy efficiency. In comparing any material with another, the real question that must be answered is, "What is the best value we get from a kilogram of petroleum?"

Whenever SMC is compared to metals for exterior body parts, the question of relative energy efficiency often arises. Fig. 14 shows how favorably SMC compares in energy savings over a four year life period of an SMC car hood

	Weight kilograms	Total energy to produce m kilo calories	Gasoline energy during car life m kilo calories	Total energy usage m kilo calories	Percent energy savings versus Steel
Steel	34	529	2126	2655	Base
Aluminium	17	1002	1049	2051	23 %
2-Piece SMC	21	475	1332	1807	32 %

Fig. 14. Total energy used during manufacture & lifetime of intermediate-sized hood.

vs. the same hood made from steel and aluminium. It should be noted that also included in the energy to produce the SMC hood is the alternative BTU value of the SMC itself if it were to be burned for fuel. This chart was developed several years ago by Owens Corning Fiberglas based on conventional SMC. With tougher, more flexible SMCs the two piece hood could be made even thinner and therefore still less energy consuming.

Most materials such as aluminium, light weight steels, and other plastics will all have a place in the race for weight reduction of automobiles. However, with continuing development, such as the tougher, more flexible SMC as described, SMC can and will be the growth leader for exterior plastic automotive body panels.

Polycarbonate for automotive exterior applications

Conrad Zumhagen

Polycarbonate is a family of high performance amorphous thermoplastics developed and manufactured by General Electric Company, Plastics Division. This paper will discuss some of the applications of these materials for automotive exterior components.

Lighting

Because of its combination of high impact, high heat, and transparency polycarbonate is ideally suited for automotive lamp components. The polycarbonate system was developed during the 1960's to meet U.S. agency in 1969, U.S. car and truck manufacturers began utilising Polycarbonate in a variety of car and truck signal lamps for improved performance in the areas of heat, impact, durability and design freedom. Today, Polycarbonate lamp lenses and/or housings are commonly used in significant numbers in the US on every type of signal lamp and on all types of cars, trucks and buses. Prior to the 1980 U.S. model year, the only vehicle lamp not manufactured in Polycarbonate (or orther plastic) was the headlamp.

First plastic headlamp developed for production

It has been felt for many years that plastic headlamps could offer many advantages over glass – most importantly weight reduction and reduced breakage. Until recently, U.S. headlamps construction has been of the "sealed-beam" type, with a Tungsten filament in an inert atmosphere enclosed by a hermetically sealed glass lens and reflector/housing. The gas permeability of plastics vs glass prevented plastics' usage in this device. The recent adoption of the encapsulated halogen light source for U.S. headlamps has solved this problem and opened the door for plastic headlamps development.

General Electric Company Miniature Lamp Products Dept. began a project to develop and demonstrate production feasibility of a sealed-beam plastic headlamp. The lamp type chosen for direct replacement was the current

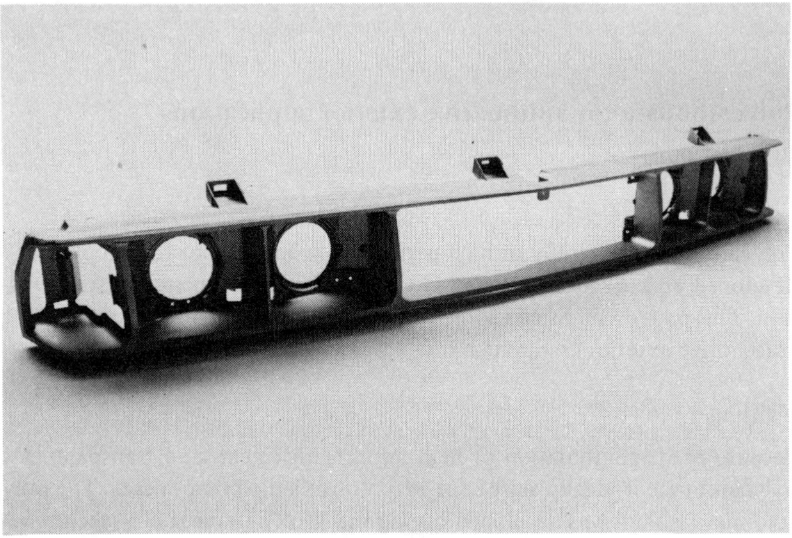

Fig. 1. Peugeot 604 Federal Model in NORYL PX1112. Fully integrated headlamp and indicator housing.

production 165MM rectangular (high beam pair of a 4-lamp system) glass unit.

After extensive evaluation of all plastic candidates for both lens and reflection in prototypes, Polycarbonate was chosen on the basis of performance, manufacturing and economics. The construction of the unit is the following:

Injection moulded, coated Polycarbonate lens.

Injection moulded, metallized Polycarbonate reflector.

55 watt encapsulated halogen light source mounted in Polycarbonate black and ultrasonically welded into reflector.

Lens and reflector welded together in sealed unit by linear/vibration process.

Advantages of new headlamp

The new Polycarbonate headlamp (designed type 1A1, GE 4HP 4651) as developed offered a number of significant advantages:

1. 68 % weight reduction vs current glass unit (.18 kgs vs .57 kgs).

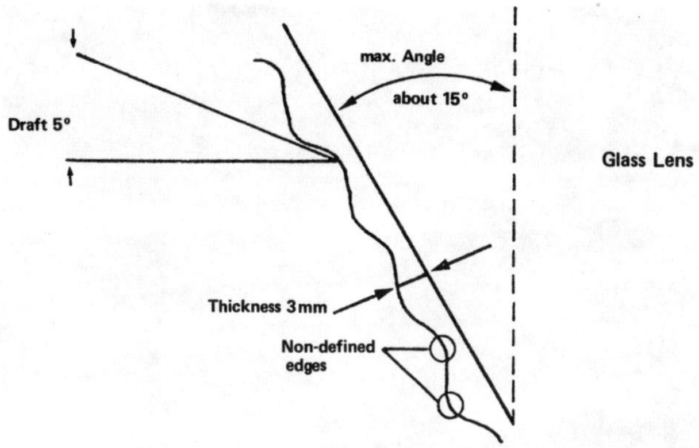

Characteristics	Glass	PC
Wallthickness	5 - 6 mm	3 - 4 mm
Specific weight	2.4 - 2.7	1.2
Draft angle for ejection	5°	0.5°
Tolerances	wide	narrow
Maximum declination of lens	15°	>45°
Lifetime of tooling	ca. 1000 shots	500.000 shots

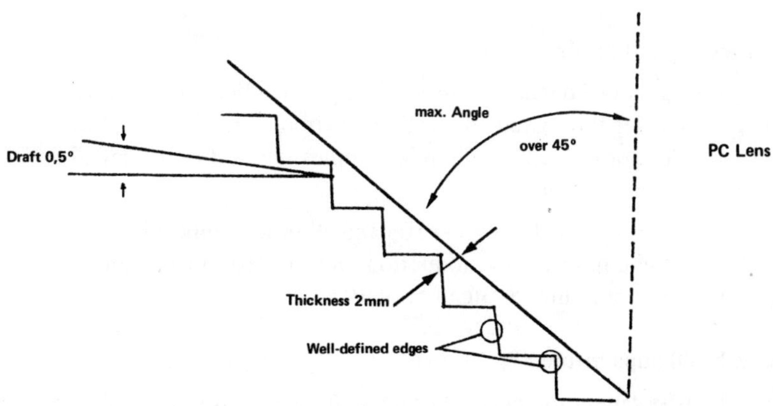

Fig. 2. Characteristics of Glass Lens and PC Lens.

Fig. 3. Fiat 131 Racing – front and rear bumper plus wheel arch extenders in LEXAN ML3026.

2. Better optical performance. Plastic injection moulding yields more accurate and reproducible optics and beam pattern control, resulting in a 20 – 30 % improvement over glass.
3. Safety: Polycarbonate is virtually unbreakable whereas glass headlamps are easily and frequently broken in use.

Field testing and production

Production quality Polycarbonate headlamps have been extensively field tested on vehicles throughout the U.S. in environment of – 40 °C to 50 °C. After over 20 months and 1 million lamp miles, the test units met all SAE photometric requirements.

The current lamp meets FMVSS – 108 and all other applicable U.S. agency requirements and has been in production on Ford Motor Company's 1980 Lincoln Continental since September of 1979.

Plastic headlamps future

While the first generation plastic headlamp has been proven in Polycarbonate its design is basically a direct replacement of a glass unit and does not take advantage of many of the further possibilities of plastics proven on many other parts. New plastic headlamp designs incorporate such features as integral attachment/aiming functions to optimize cost/weight, and improved aerodynamics/styling.

Work done to date shows Polycarbonate to be an excellent headlamp material for international applications.

European story – headlamps

Headlamps construction in Europe is as is well known completely different to the USA. In principle today European lamps are much larger and complex giving at least double the light output and hence are much hotter. This radically different construction and styling together with different legislation both provides an opportunity and constraints for the adoption of plastic materials.

The use of plastic materials for the reflector and housing is becoming well established especially in France where the very style of the lamps beg for plastics construction. The progression from plastic extenders to reflectors has moved in step with raw material innovation. High temperature grades of modified PPO's have played a major part in the headlamp programs of Cibie, SEV Marchal, Seima and Ducellier.

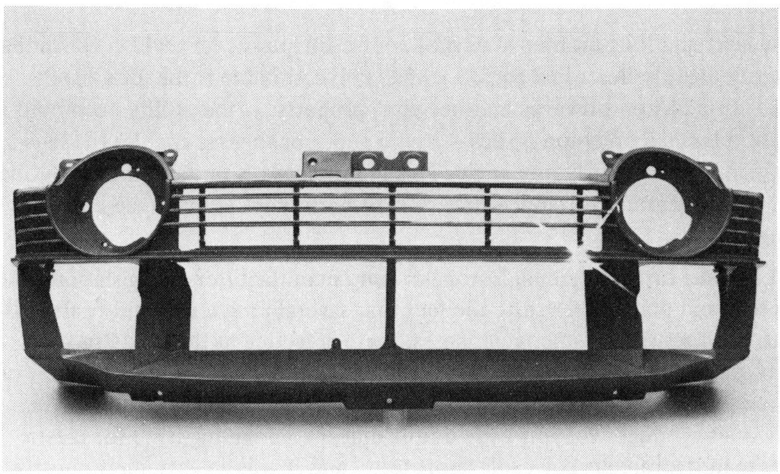

Fig. 4. The Fiat Strada front end in LEXAN ML3026. Produced in twin impression tool together with the rear end by AG International in Torino.

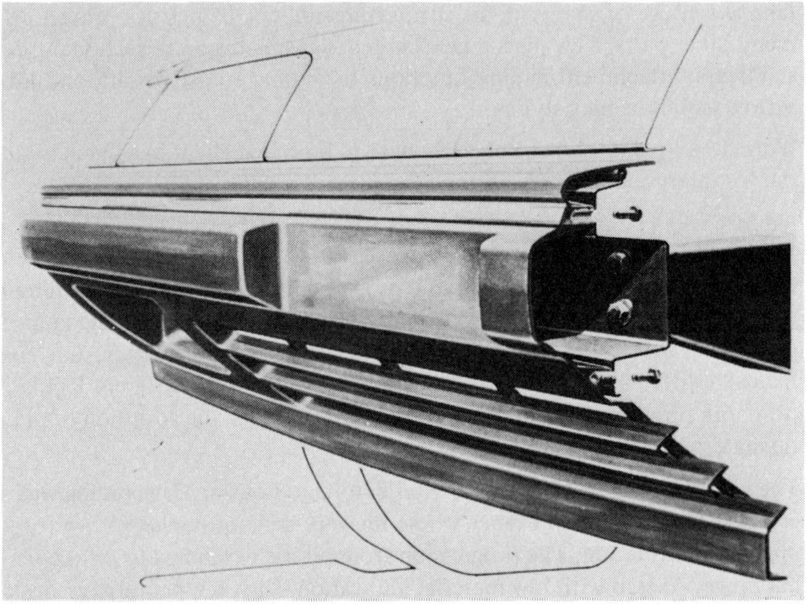

Fig 5. Typical concept for LEXAN polycarbonate bumper system using rolled steel support for impact up to 4 kph.

The next and long awaited step is of course the plastic lens which for similar reasons already described for US lamps, Polycarbonate is the ideal candidate. In addition however another vital property – the ability to mould close tolerance precision optics – by far superior to glass enables the lens to be inclined at virtually any angle. This requirement is pushed heavily by the car manufacturers to reduce drag and to harmonise styling (see enclosed diagram).

A key issue however facing European lamp manufacturers is legislation both present and planned. Despite the fact that several long term field trials indicate good results for uncoated lenses, there is feeling in the industry that only glass like surface hardness will suffice. Consequently a large number of abrasion resistant coatings are under evaluation including SHC1000 a GE Silicones product. There is little doubt that the remaining issue for plastic lens manufacture are solveable short term, and it will be extremely surprising if a Polycarbonate lens does not appear on full scale production vehicle before the end of 1980.

Exterior black trim

As the exterior trim in the U.S. has been a trend towards black versus bright polycarbonate has been chosen for a number of applications because of its excellent mechanical properties over a wide range of temperatures. Polycarbonate's good U.V. resistance and colour stability have allowed it to be used with moulded – in colour, offering improved performance, while being competitive with other plastics that must be painted.

Some examples of U.S. production applications are:

Chevrolet "B" body rear window mouldings

GM "X" body front window "plug"

GM "X"-body B-pillar cover

Buick "H" body (Skylark) B-pillar cover.

Bumpers and front ends

Polycarbonate got its start in the bumper/frontend area in components of the 1968 and 1969 Pontiac Firebird. This car (produced before the advent of U.S. bumper standard) used injection moulded bodycolour painted Polycarbonate in the headlamp bezels and grille surrounds subject to frontal impact. These vehicles, still on the road today have demonstrated the durability of painted Polycarbonate in the U.S. environment for over 10 yrs.

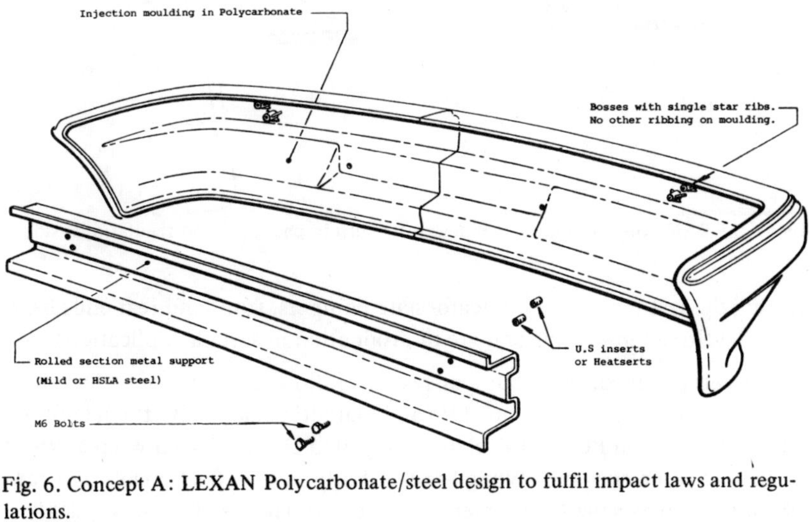

Fig. 6. Concept A: LEXAN Polycarbonate/steel design to fulfil impact laws and regulations.

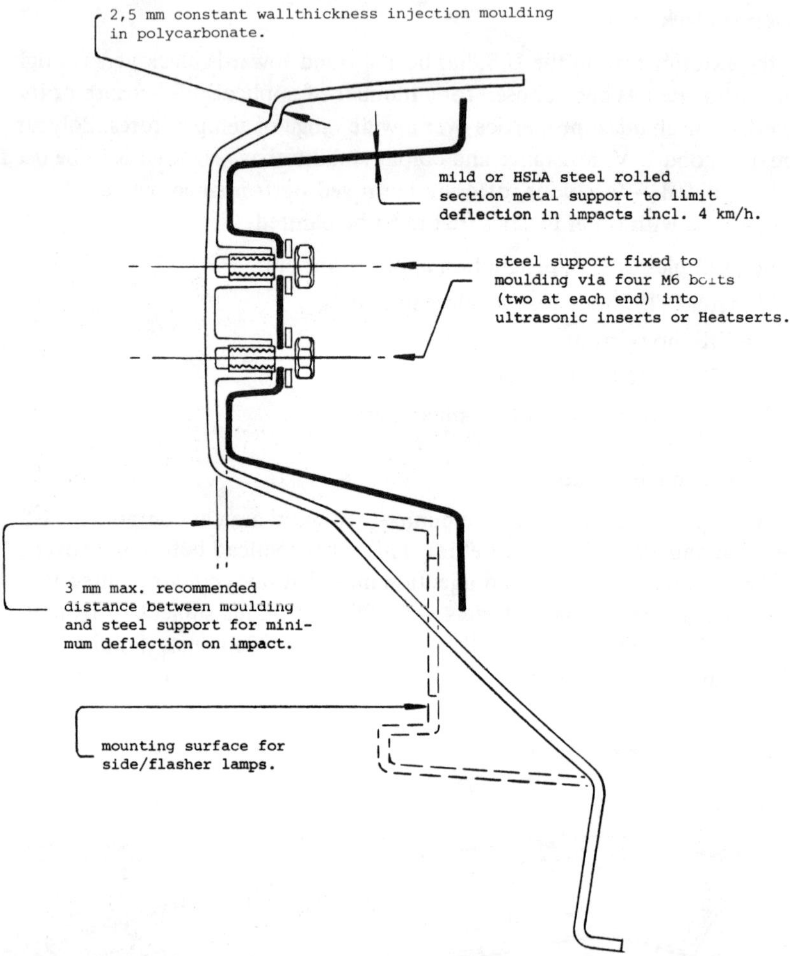

Fig. 7. LEXAN Polycarbonate/steel design to fulfil impact laws and regulations: side-view.

During the last few years, Polycarbonate bumpers, front and rear ends have been developed in Europe and are in production in several applications.

The development of these bumper systems followed a series of logical steps in tune with rising production costs and market demands. By the late 60's plastic end caps in PU and PP related integrated chromed steel wrap arounds, to achieve some cost and styling benefit. Early in the 70's Renault pioneered SMC bumper shields mounted over steel armatures. This was quickly

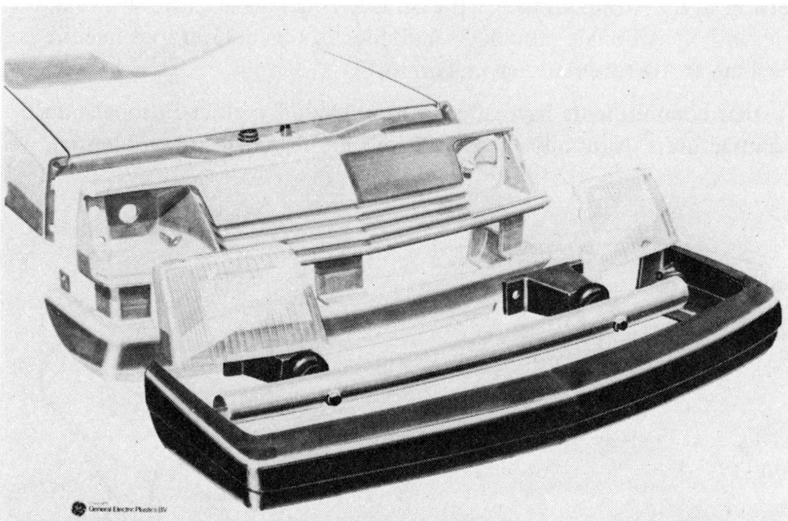

Fig. 8. Proposal for front end bumper assembly to accomodate 4 or 8 kph. The front end integrates grill headlamp houses and spoiler: whilst the bumper is mounted separately free to move on minor impact.

followed by VW and Fiat who independently developed different systems based on EPDM/PP. Even more recently Fiat introduced on the Ritmo a self supporting front end in EPDM/PP, whilst this year Citroen introduced their unusual two colour EPDM/PP unit backed up by parallel PP tubes.

Clearly then we see in Europe a great deal of very creative innovation which unlike the US is unfettered by unwieldy legislation – at least for the present. The overall trend has been for much larger complex bumper systems that integrate several functions on both vehicle protection, assembly and styling. This revolutionary thinking has created a unique demand for high performance engineering plastics – and quite suddenly Polycarbonate found itself a favoured material for these fast changing new market requirements.

The unique balance of properties Polycarbonate provides made it the ideal candidate for several recent European models – first the BL Triumph TR7, followed by the Fiat 131 Racing which illustrates the dramatic styling opportunities available with Polycarbonate. As we shall see the 131's construction points the way to what we consider could be the optimum solution to meet the forthcoming ECE requirements.

In addition to bumpers Fiat also introduced Polycarbonate on the federal version of the Ritmo known as the Strada. This vehicle carries front and rear ends in Polycarbonate moulded simultaneously on a 3500 tons injection machine at AG International in Torino.

Further commitments have already been made by other European automanufacturers. Naturally these decisions are regarded as confidential, we

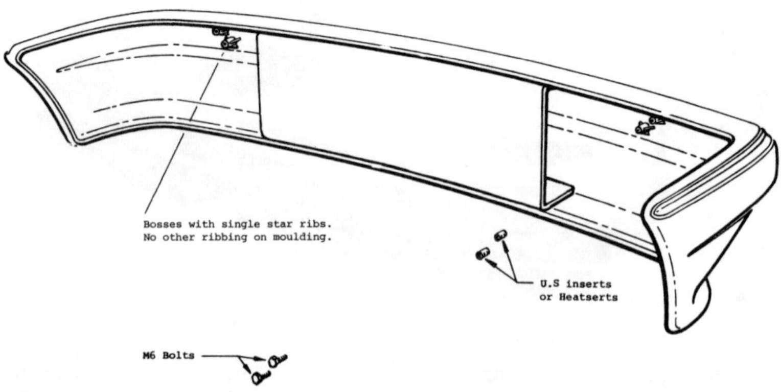

Fig. 9. The Concept B: LEXAN polycarbonate bumper: performing up to 6 – 8 kph.

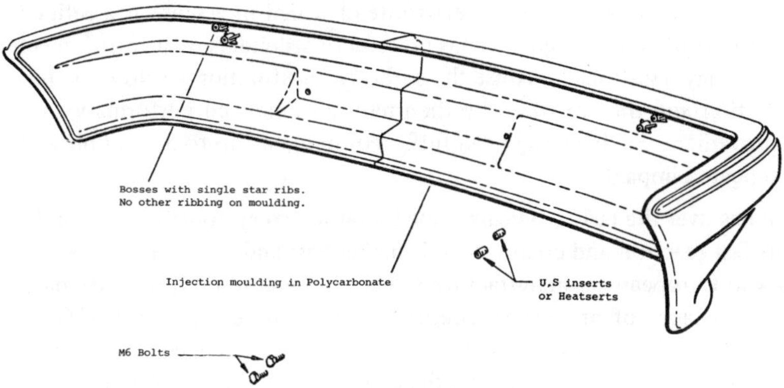

Fig. 10. Concept C: Cosmetic design: can perform very well up to 3 – 4 kph.

can however announce that during 1980 we shall see the launch of at least three other vehicles carrying Polycarbonate bumpers:

1. The Lancia Beta Monte Carlo replacing a duplo PU/ABS system.
2. The Innocenti Milli.
3. The Zastava 102 from Yugoslavia – a bumper designed from the outset to meet 4 km/h since Yugoslavia has elected already to adopt the ECE21 draft legislation.

Future developments will follow three GENERAL ELECTRIC PLASTICS established concepts.

A) Polycarbonate cladding fixed to a rolled steel section or simple stamping. The steel profile may be very light and inexpensive. Its function is to assist in the energy management of the impact. It can be designed to withstand 4, 6 or 8 kph with plastic or elastic deformation dependent on requirements. The use of HLS steels will play a major role in such systems of the future. The Polycarbonate moulding is very thin 2.5 mm and generally does not require more than two fixing points on small vehicles. Four may be required for larger cars with long wrap arounds or partial front ends. This simple Polycarbonate moulding has a mould cycle time as low as 60 secs. for a pair of bumpers, thus making it an extremely competitive solution against traditionally used materials.

B) Closed Polycarbonate bumper fulfilling legal requirements up to 8 kph without the use of steel re-enforcement – as the USA Fiesta test demonstrates.

C) The all Polycarbonate bumper/frontend which in a sense is a refined form of B). This design also has no steel or secondary rear re-enforcement – it may additionally utilise the carbody construction to close the box section and thus provide the then naturally induced performance improvement. Such a design can fulfil requirements up to 3 – 4 kph with repeated impact.

GEP has over the last two years gained a rather exceptional reputation for their fast reaction and creativity in bringing cost and weight effective solutions to European car manufacturers. The combination of polycarbonate's unique balance of properties, coupled with the total capability of GEP's resources worldwide offers the auto industry an unparallelled opportunity for mutual co-operation in facing the challenges of the next decade.

Modified polycarbonates developed

Using recent technology developed by General Electric Plastics in the U.S. and Europe, new, modified grades of Polycarbonate have been produced for bumpers and other energy-management automotive applications. Polycarbonate as we saw earlier has found a natural fit keyed to the European vehicle requirements:

Exceptional dimensional stability and creep resistance.

Superior impact over wide temperature range.

High rigidity capable of being self supporting in thin sections.

Fast mouldability – 60 parts per hour.

Wide design and styling freedom – including colours.

High temperature resistance to meet retouch oven requirements.

Easy paintability without expensive pre etching processes.

Plans are at an advanced stage for the introduction of yet a third generation of Polycarbonate grades which will fulfil the increasing market demand for still further improved petrol resistance.

In the U.S., a new family of modified polycarbonates has been developed. EM grades offer significantly improved impact at low temperatures and in thick sections. Those materials (currently in production usage for the instrument panels in Ford Motors Company's 1980 Thunderbird, Cougar and A-series trucks) provide high performance over a wide range of temperatures and increased ability for bumper design flexibility.

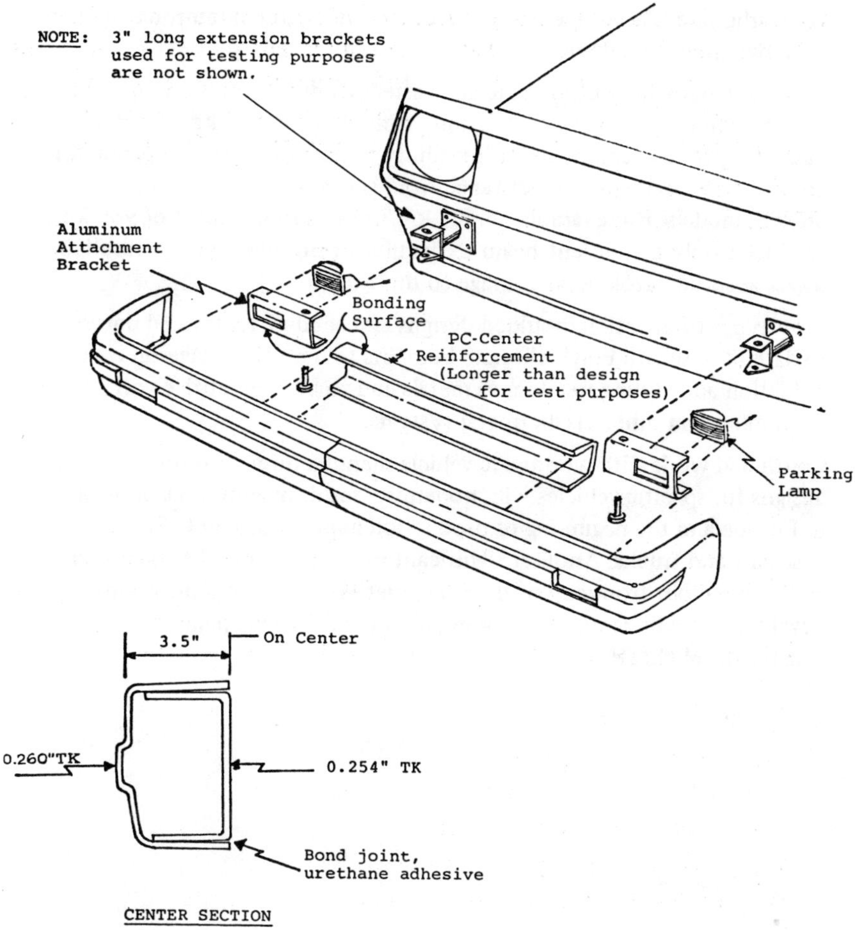

Fig. 11. Exploded view of LEXAN Polycarbonate bumper system, as tested.

U.S. bumper development

With the history of European bumper production and new Polycarbonate technology, GE Plastics has been developing a lightweight bumper to meet U.S. requirements (FMVSS part 581 8 km/h frontal impact: 4.8 km/hr 30° corner impact, with no damage). A basic design has been developed for a

Polycarbonate EM bumper system requiring no material reinforcement and only two mounting points.

Prototype parts have been made and tested pendulum, plus 8 km/hr. barrier to U.S. standards on vehicles weighing from 910 to 1315 kg. The Polycarbonate bumper has demonstrated the ability to manage energy in pendulum impacts with and without assistance from shock absorbers commonly used on U.S. models. For example, an 8,1 km/hr centerline impact of 962 kg produced only a transient beam deflection of 36 mm with no resultant shock absorber stroking or damage to the system.

Prototypes of injection moulded designs and solid-phase formed parts tested to date have shown Polycarbonate bumpers to have advantages in weight reduction and styling freedom vs metals and to be competitive with current aluminium and other plastic/metal systems.

Continued work with worldwide vehicle manufacturers toward optimizing designs for specific vehicles will, hopefully, result in additional applications in Europe and the beginning of production applications on U.S. and Japanese cars and trucks. Another significant production use of Polycarbonate is the body colour front end of the Chevrolet Monza. This application and development will be discussed in detail in the following paper by my colleague, Mr. Michael West.

Polycarbonate for automotive front ends

Michael W. West

Polycarbonate for automotive front ends

In selecting the proper material for automotive front ends, consideration must first be given to the requirements of the part. In the case of a front end, the automotive engineer is looking for a material that is light weight, rigid, has low creep (especially if the headlamp housings are included) and demonstrates good dimensional stability, chemical resistance, and impact. If the front end is an appearance part, the material must provide a finish suitable for painting.

The material must be able to be moulded accurately into the size and shape of the part and provide part-to-part consistency. Finally, when all of these factors are evaluated, the completed part in the material has to be cost effective.

The material also must have good thermal resistance to withstand paint baking where required, as well as U.S. environment of 50 °C, engine and headlamp heat.

A material which meets all of these requirements and has a history of use in the headlamp housings and front ends is 10 % glass filled polycarbonate. Let us address each of the issues.

Light weight

10 % glass filled Polycarbonate has a specific gravity of 1.25 g/cm^3 which favourably places it on the low end of the spectrum of thermoplastic and thermoset materials suitable for front end components.

The physical and processing properties of the material enable the front end to have a wall thickness of only .110" (2.8 mm) which, again, contributes toward weight savings.

Eliminating or combining parts contribute towards weight savings. The low creep characteristics of 10 % glass filled Polycarbonate enable the headlamp housings to be incorporated into the front end. This eliminates 2 lbs. (.9 kg) and several fasteners from the vehicle.

Rigidity

The concern with rigidity is that the material should not deform between fixturing points. 10 % glass filled Polycarbonate has a flexural modulus of 500000 psi (3700 N/mm^2) which has proven to be satisfactory.

Proper ribbing can be selectively incorporated into the part to add reinforcement where necessary. This is essentially a cost-free method of adding rigidity to the part.

Headlamp aim

Polycarbonate exhibits excellent creep characteristics (see fig. 1). It has been successfully used in numerous headlamp housing applications.

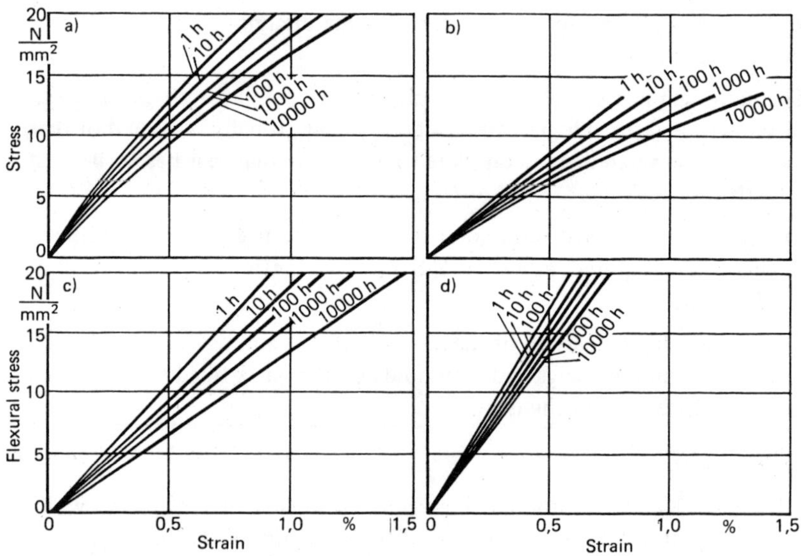

Fig. 1. Creep characteristics of polycarbonate.

Dimensional stability

Heat generated by the radiator and the sun can cause concern. However, glass filled Polycarbonate, having a heat deflection temperature of 145 °C (1.82 N/mm^2) overcomes these conditions.

The coefficient of thermal expansion being 4 x 10^{-5} m/m °C, minimizes the concern over deformation between fixturing points and maintains overall part stability.

The low hygroscopic nature of Polycarbonate also helps to maintain overall part stability. The equilibrium water absorption value for 10 % glass filled Polycarbonate is 0,31 % (at 23 °C), therefore there is negligible dimensionable change with Polycarbonate due to humidity changes.

Furthermore Polycarbonate being an amorphous material ensures accurate moulded shape and part-to-part dimensional consistency. Warpage due to differential shrinkage of crystalline materials is not a problem with Polycarbonate.

Chemical resistance

If the front end panel is an exposed part it can come in contact with engine and transmission fluids. Whether the part is exposed or consealed it will come in contact with road contaminants and salt.

Polycarbonate has a long history (8 years) of applications in handling housings, sidemarkers, park and turn and tail lamp lenses and reflectors. With the exception of reflective optics lens applications, coatings are seldom used on these parts. The widespread and long term use of these applications have not uncovered any chemical compatibility problems with the application of Polycarbonate on a vehicle front end.

Where the front end panel will be exposed and painted body colour, the primer also performs as a barrier coat. The primer, being cross-linking not only acts as a means of accepting the topcoat, but also provides a barrier to many hostile chemical agents.

Impact

The concern over impact comes from 2 areas. The first is normal car use: stone impingement, dropped items, cleaning, ice removal, etc. The second concern occurs earlier: at the assembly plant. The part has to withstand normal use such as stacking and fixturing, but it may also be dropped or banged about during transport and assembly procedures. The excellent impact characteristics of Polycarbonate (see fig. 2) make it an optimal material because of these concerns.

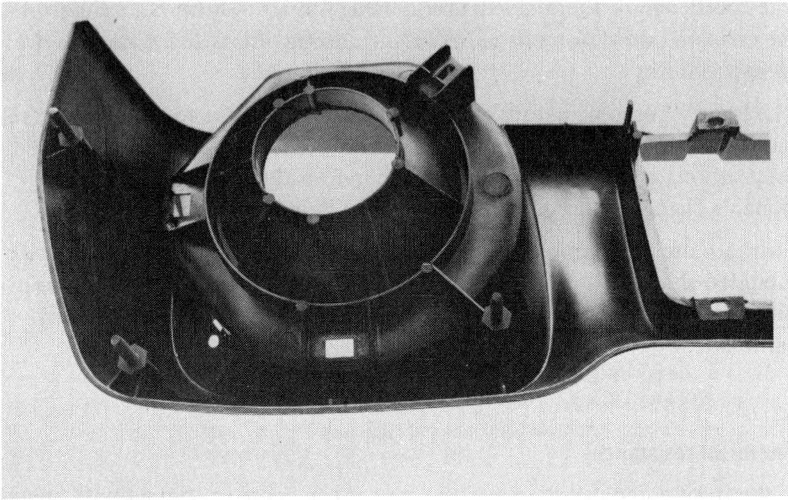

Fig. 2. Chevrolet Monza in Polycarbonate showing integration of headlamp housing capable of fulfilling strongest SAE requirements.

Manufacturing considerations

Injection moulding large components in Polycarbonate involves many key critical considerations. Because of the large complex sections, several gate locations are usually needed, for the same reason, the material flow must be carefully planned to 1. optimise surface appearance and 2. assure proper filling.

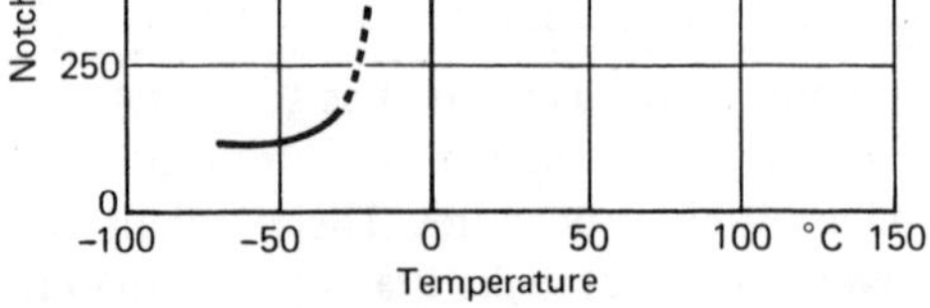

Fig. 3. Notched impact strength vs temperature.

To meet these needs, special attention must be paid to manifold design temperature control zones and nozzle configuration. The general recommendations below apply to all large components. Some special requirements may be needed with unusual configurations.

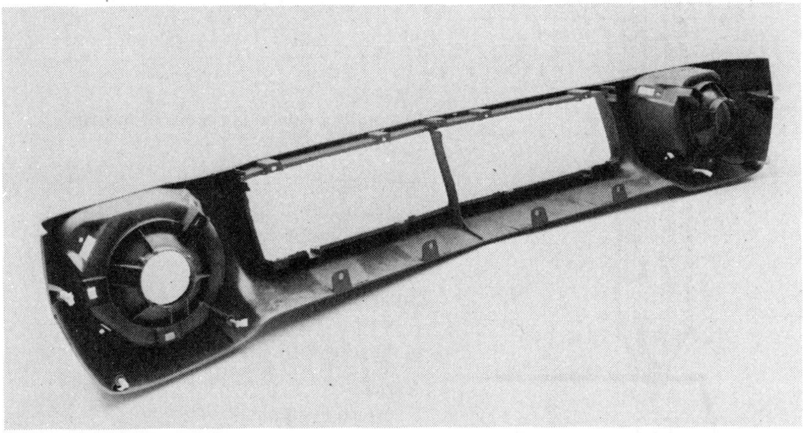

Fig. 4. Chevrolet Monza Town Coupé in Polycarbonate rear view showing mounting integration features.

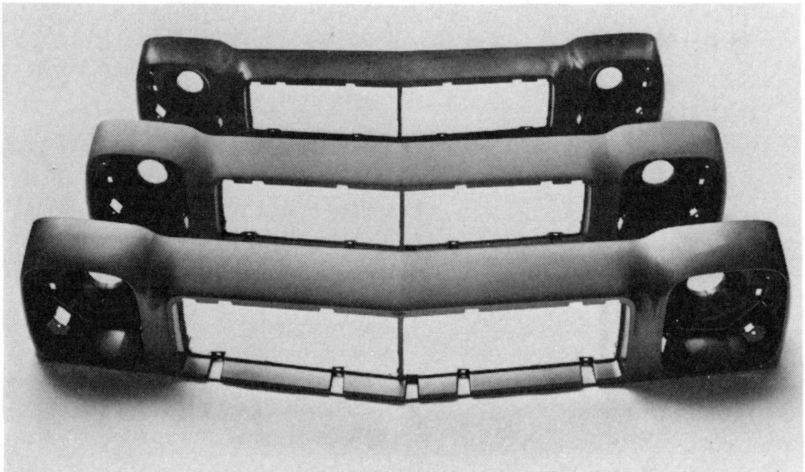

Fig. 5. Chevrolet Monza Town Coupé in Polycarbonate. Three stages of production top of bottom: 1. as moulded; 2. primed at moulder, 3. finally with PU topcoat.

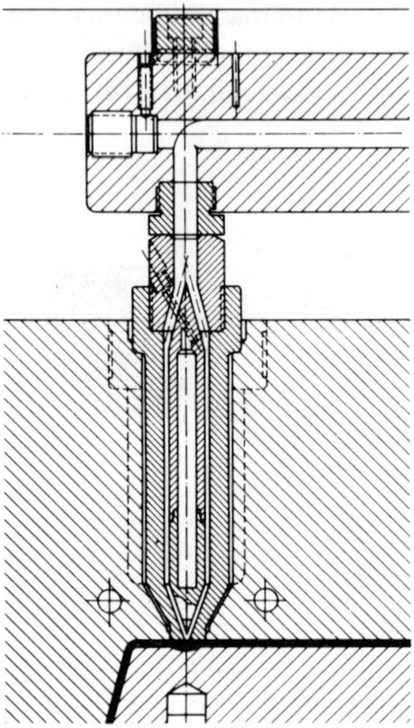

Fig. 6a.
Internally heated 'Torpedo' or 'hottip' bushing systems.

Advantages:
- readily available
- siutable for most tooling
- slide sealing

Disadvantages:
- vulnerable system
- heating difficult to control

Advice:
- consult 'suppliers' before starting tool design

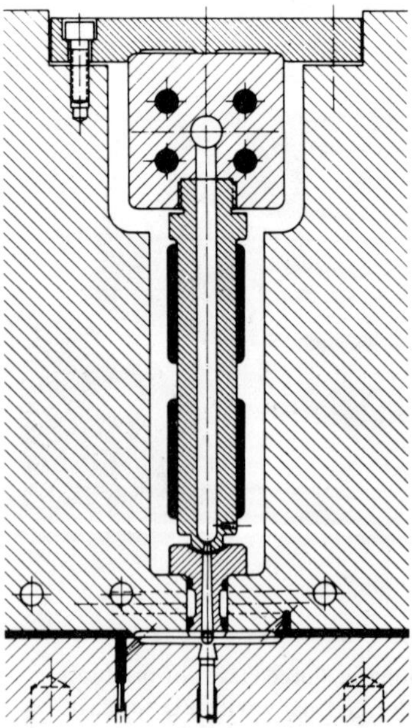

Fig. 6b.
Externally heated 'open' nozzle system combined with a cold sprue bushing.

Advantages:
- low manufacturing costs
- easy to operate
- easy to control
- suitable for GE plastics
- suitable for almost all hotrunner systems

Disadvantages:
- not suitable for extreme long manifold blocks due to stiff connection to the block (linear expansion)

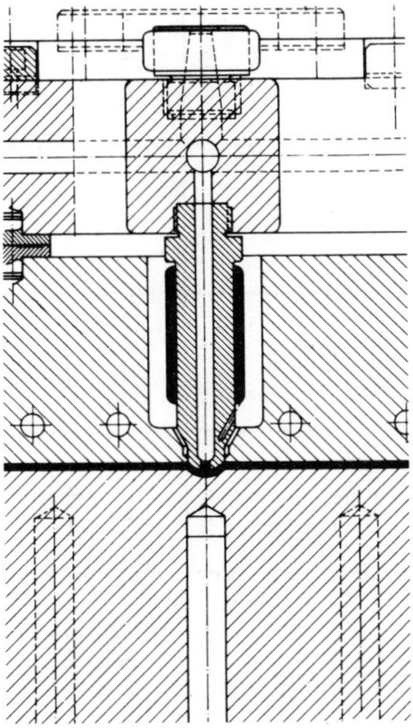

Fig. 6c.
Externally heated 'open' nozzle system with pin-point.

Advantages:
- lowest manufacturing costs
- easy to operate
- easy to control
- suitable for GE plastics
- suitable for almost all hotrunner systems

Disadvantages:
- not suitable for extreme long manifold blocks due to stiff connection to the block (linear expansion)

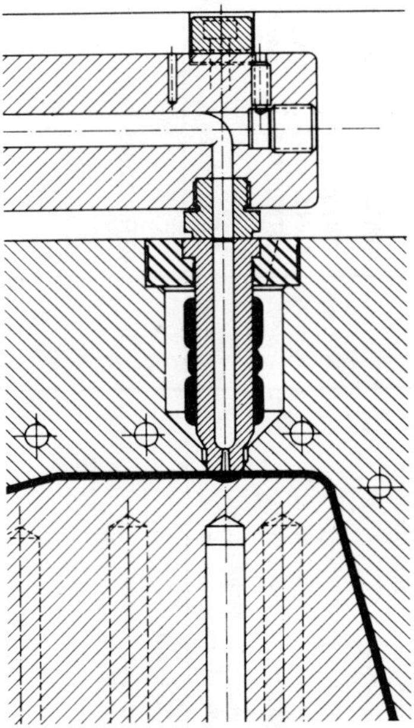

Fig. 6d.
Externally heated 'open' nozzle system with pin-point sliding sealing.

Advantages:
- easy to operate
- easy to control
- low manufacturing costs
- suitable for all GE plastics
- suitable for 'all' hotrunner systems due to 'sliding' sealing

Disadvantages:
- none if precisely made

Properties	Unit	Testmethod	120, 140, 160, 100 series	2014, 2034	500	3412	3414	940
Mechanical								
Tensile strength at yield	N/mm^2	ASTM D 638	>60	>60	65	—	—	>60
at break	N/mm^2	ASTM D 638	>70	>70	50	90	120	>70
Elongation at yield	%	ASTM D 638	6-8	6-8	4-5	—	—	6-8
at break	%	ASTM D 638	>100	>80	10-20	3-4	2-3	>100
Tensile modulus	N/mm^2	ASTM D 638	2300	2300	3500	6000	10000	2300
Flexural yield strength	N/mm^2	ASTM D 790	100	100	—	—	—	100
Flexural break strength	N/mm^2	ASTM D 790	—	—	110	145	190	—
Flexural modulus	N/mm	ASTM D 790	2500	2500	3700	5600	10000	2500
Notched impact strength Charpy	kJ/m^2	DIN 53453	>30	>30	>15	8-10	8-10	>30
Notched impact strength IZOD	J/m	ASTM D 256	600-800	600	200-300	100	100	600-800
Hardness, H 358/10	N/mm^2	DIN 53456	98	100	115	125	145	98
, H 358/60	N/mm^2	DIN 53456	93	95	110	120	140	93
, Rockwell	—	ASTM D 785	M70	M74/R120	M85	M91	M93	M70
Abrasion resistance, Taber CS 17, 1000	g	ASTM D 1044	0.010	0.009	0.011	0.017	0.032	0.010
Thermal								
Oxygen Index**)	%	ASTM D 2863	25	31	36	32	34	35-40
Flame retardancy** (1.6 mm thickness)	—	UL stand 94	94 V2	94 V2 94 V0 (2.3 mm)	94 V0	94 V1	94 V1	94 V0***)

Heat resistance:								
VST/B	°C	VDE 0302	160-170	160-170	155-165	160-170	165-175	160-170
Vicat, method B	°C	ASTM D 1525	150-155	150-155	155-160	160-165	160-165	150-155
H.D.T., 1.82 N/mm^2	°C	ASTM D 648	135-140	135-140	140-145	145-150	145-150	135-140
Mould shrinkage****)	%	ASTM D 1299	0.5-0.7	0.4-0.6	0.2-0.55	0.2-0.5	0.1-0.3	0.5-0.7
Thermal conductivity	W/m°C	ASTM C 177	0.20	0.20	0.21	0.22	0.22	0.20
Coeff. of linear thermal expansion****)	m/m°C	ASTM D 696	7×10^{-5}	7×10^{-5}	4×10^{-5}	3×10^{-5}	2×10^{-5}	7×10^{-5}
Physical:								
Specific Gravity		ASTM D 792	1.20	1.24	1.25	1.35	1.52	1.21
Water absorption								
24 hrs., 23°C	mg	ASTM D 570	10	10	9	10	9	10
equilibrium, 23°C	%	ASTM D 570	0.35	0.32	0.31	0.29	0.23	0.35
equilibrium, 100°C	%	ASTM D 570	0.58	0.58	—	—	—	0.58
Light transmission	%	ASTM D 1003	88-89	87-88	—	—	—	88
Electrical								
Volume resistivity	Ohm x m	ASTM D 257	10^{15}	10^{15}	10^{15}	10^{15}	10^{15}	10^{15}
Dielectric strength at 3.2 mm	kV/mm	ASTM D 149	15	15	18	20	20	17
Dielectric constant:								
50 Hz, 23°C		ASTM D 150	3.0	3.0	3.1	3.2	3.4	3.0
1 MHz, 23°C		ASTM D 150	2.9	2.9	3.0	3.2	3.4	2.9
Power factor: 50 Hz, 23°C	—	ASTM D 150	0.0009	0.0009	0.0009	0.0009	0.0013	0.0009
1 MHz, 23°C	—	ASTM D 150	0.010	0.010	0.0075	0.0073	0.0067	0.010
Arc resistance 23°C:								
Stainless steel	s	ASTM D 495	10-11	10-11	5-10	5	5	10
Tungsten	s	ASTM D 495	120	120	120	120	120	120

*) Typical values only. Variations within normal tolerances are possible for the various colours. **) This rating is not intended to reflect hazards presented by this or any other material under actual fire conditions. ***) At 3.2 mm for 940N clear, at 2.0 mm for 940N opaque, at 1.0 mm for 940NL, ultra thin applications. ****) For glass reinforced grades, variations are possible depending on the glass fibre orientation.

Mouldings parameters

Hot runner manifold
automatic degater
90 second cycle time gate-to-gate.

Part finishing

Primer

Applied at moulder
2 part urethane Spraylat Corp. AP9112
.75 mils thick
250 °F bake for 35 minutes.

Topcoat

Applied off-line at assembly (system in place necessary for soft trim components).
PPG Durathane 300
250 °F bake for 35 minutes.

Fixturing

6 "hi/low" studs, inserted at moulder, on the back side of the part near the ends.

"J-nuts" are applied on moulded part in flanges at the assembly plant.

The "hi/low" studs are used to center the part on the vehicle. The hi/low and J nuts combine to fixture the part to the vehicle, allowing for body sheet metal assembly tolerance.

Summary/conclusions

We have seen the evolution of parts using glass-filled Polycarbonate from discrete headlamp housings to complex headlamp housings/cross-over panels requiring a "Class A" body colour finish. The development of new high impact energy management grades in the U.S. and Europe brings interesting new possibilities into front end design where bumper fillers and/or total energy management systems can be combined with Monza type front ends.

Automotive applications of rim body panels, fascia and foam bumper systems

Peter A. Weller

Introduction

The ever increasing pressures of fuel shortages, rising fuel costs and government regulations have greatly accelerated the use of plastics in automobiles. Polyurethane products have been a leading participant in this growth, and they are expected to continue to provide low weight, low cost answers to the automotive need for improved materials and reduced weight.

The purpose of this report is to describe some of the applications in which polyurethane has been, is, and will be used to considerable advantages on automotive exteriors.

Past and present bumpers and fascia

Up to the present day, bumpers have been the principal exterior component where urethane has been used in significant quantities. Even before there were any bumper standards such as Standard 215 in the United States, urethane faced bumpers that had significant damage resistance in parking lot impacts were being produced. These were simply a layer of dense, but resilient, urethane foam, molded directly to a metal bumper bar and painted body color with a flexible paint. They were hard, heavy, and relatively expensive and were used exclusively on sports cars. With the advent of MVSS Standard 215, the foam was thickened and made softer by lowering the density. Energy of the 5 mph impact was absorbed by compressing the foamed urethane.

To reduce the weight, the trend in Europe was to core out the foam to form a series of high density ribs which absorb energy by collapsing under impact, thereby, maintaining the integration of the cosmetic outer surface and the energy absorber as a single component. This is commonly known as the HIFLEX design and was used on the Ford Capri, the MG and the Volkwagen Dasher.

In the States, we chose to separate the skin from the energy absorbing foam so that [1] the foam could be formulated for maximum energy absorption

Table 1. Design alternatives.

Hiflex: Integrated functions

Thin fascia: Separated functions, optimized processing; cost

DESIGN	WEIGHT (KG)	CAR WEIGHT (KG)
CAPRI OR VW HIFLEX	7-8	1200
CAPRI THIN FASCIA & SEPARATE E/A ESTIMATE	4.5 - 5	1200
1979 FIREBIRD FULL THIN FASCIA & SEPARATE E/A	6.4	1800

efficiency at the lowest density without the need to form an integral cosmetic skin, and [2] the skin material could be formulated to produce a superior cosmetic surface at the highest production rate and lowest weight and cost. In addition, the separation of the skin from the foam made it possible to selectively size and locate the foam to meet both the requirements of the standard and the real world with the lowest volume, weight, and cost. The weight effectiveness of the separation of skin and energy absorber is shown in Table 1. Note that the Capri with a thin RIM blade type skin and a separate low density foam energy absorber would have weighed considerably less than the HIFLEX bumper. Furthermore, even on the 1979 Firebird, a much heavier vehicle with a large full fascia, the separate skin and foam together weigh significantly less than the Capri HIFLEX.

Table 2. RIM materials.

	MC-1071	MC-1073
DENSITY: (M/CM^3)	0.95	0.95
THICKNESS: (MM)	3-4	2.5
FLEXURAL MODULUS: (MPA)	180	400
TEAR STRENGTH: (KG/CM)	70	100
SAG, 120°C: (MM)	12	7
FLEXIBILITY*, -30°C:		
DURETHANE -300	PASS	PASS
DURETHANE -100	PASS	PASS

To meet the molding and performance requirements of the separate thin skin fascia, a family of RIM formulations has been developed which are low cost and have improved performance (Table 2). The latest of these materials MC-1073, was introduced into production in 1979. Even though this high performance, elastomeric RIM material (HPE-RIM) has a high flexural modulus, it is not notch sensitive at – 30 °C when painted with Durethane 100. As a result, density and thickness can be reduced to a minimum such as 2.5 mm or less.

In addition, a family of urethane foams has been developed to provide efficient energy absorption at low density. They have been formulated to have minimal changes in stiffness with changes in temperature so that they will perform satisfactorily at both high and low temperatures. Specifically, these have attainable efficiencies up to 80 % with a temperature index of 2 to 2 1/2 where the temperature index is the stiffness at low temperature divided by the stiffness at high temperature.

As a result of these efforts, RIM urethane fascia for both front and rear bumpers are now used on 25 % of all cars produced in the United States and

Table 3. Market information.
Elastomeric bumpers and fascias
Share of total bumper market

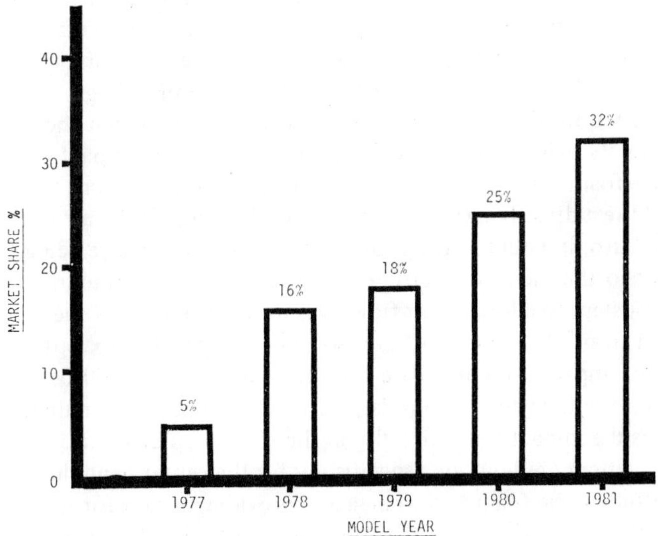

in the 1981 model year will be used on one third of all cars produced (Table 3) in the States. Of these flexible bumpers, some use urethane foam for energy absorption behind the fascia. Other use a metal bar mounted on stroking hydraulic energy absorbers behind the fascia.

The next generation soft bumper

Now in the United States, the provision in the new standard that contact with planes A and B is allowed is significant to soft bumpers because it removes the bias of an arbitrary restriction on the stroke of deflection of soft bumpers. By making the energy absorbing foam much deeper and softer, the stroke can be increased, the force reduced, and the strength of the structure reduced, thereby, reducing the weight and cost of the structure and all the attendant parts. The combination of low force and the distribution of that force by the deep soft foam as a pressure make it possible to eliminate the beam and mount the foam directly to appropriately designed and somewhat reinforced body structure such as the fenders and the radiator support structure in front, or the fenders and the rear close-out panels in the rear.

This, we believe, will be the next generation of soft bumpers. Furthermore, we believe that this concept will not only find growing application in the States but that proper evaluation and understanding by European auto manufacturers will result in significant application in the European market place. To demonstrate the function and merit of an ultra soft bumper without a beam, we decided to modify a 1978 Firebird to this design. The first problem to be overcome in this application was the fact that the radiator was directly behind the central portion of the bumper impact zone. Therefore, if the foam were to be applied completely across the front of the vehicle, it would rest directly against the radiator, blocking off the air flow to the radiator. Also, it would result in the impact force or pressure being applied directly to the radiator in either a barrier or center pendulum impact. The first step, to allow for air flow, was to remove some of the foam directly in front of the radiator. This solved every problem except a center pendulum impact, in which case the foam that remains will be deflected rearward with virtually no resistance until it contacts the radiator and still imposes the impact force directly on the radiator. The solution was to straddle the center cored-out area specifically for the center pendulum impact by reinforcing the foam in that area and providing means for transfer of the loads to the foam on either side of the cored-out area. This was done by molding into the forward leading edge of the foam a reinforcement extending across the center and into the foam on either side of the cored-out area.

To make the conversion, the conventional beam, brackets and foam energy absorber shown in Figure 1 were replaced with the one-piece reinforced deep foam energy absorber shown in Figure 2. Molded integrally in the foam is the reinforcement or "flex straddle"[1]) shown in Figure 3. Since the stock vehicle was designed with stub frames for local attachment of the brackets to mount the beam, the stock radiator support structure shown at the top of Figure 4 was not suitable for mounting the large rear surfaces of the deep foam energy absorbers. Therefore, the radiator support was rebuilt by removing the inappropriate metal, as shown at the center of Figure 4, and replacing it as shown at the bottom of Figure 4 with very strong, but light, Martinsite steel panels as shown in Figure 5. The stock fascia was

1) Patent applied for.

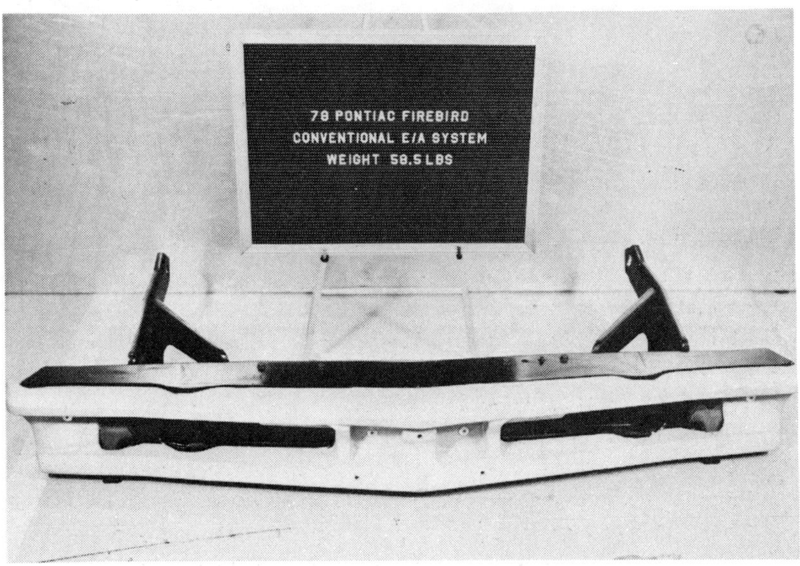

Fig. 1. The conventional beamsystem of the Pontiac Firebird of 1978.

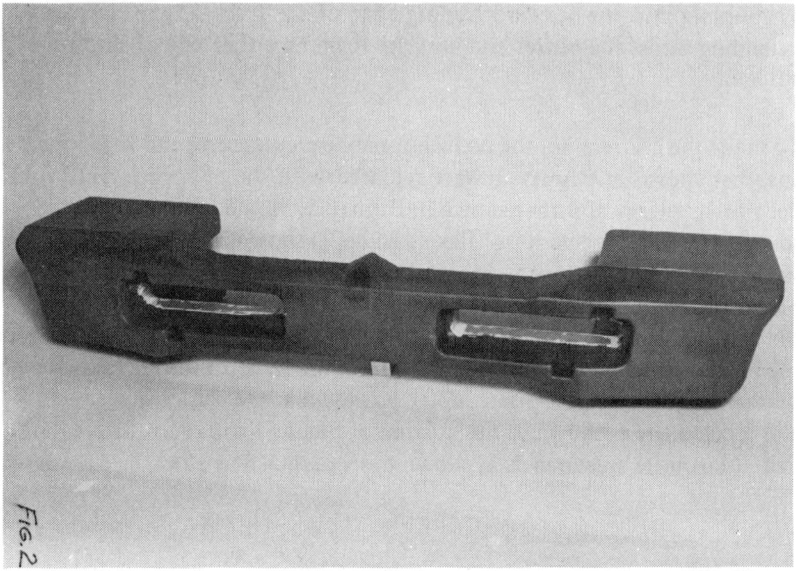

Fig. 2. The one-piece reinforced deep foam energy absorber.

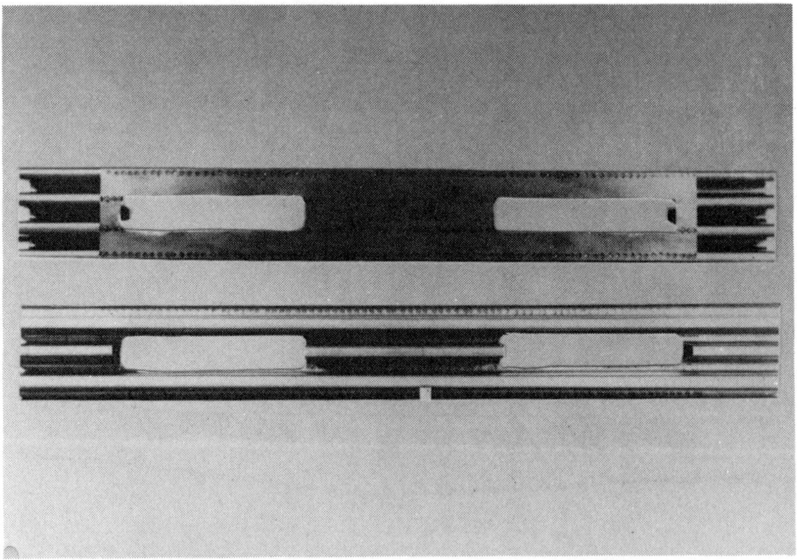

Fig. 3. Flexible reinforcement in the foam or integration of „Flex Straddle".

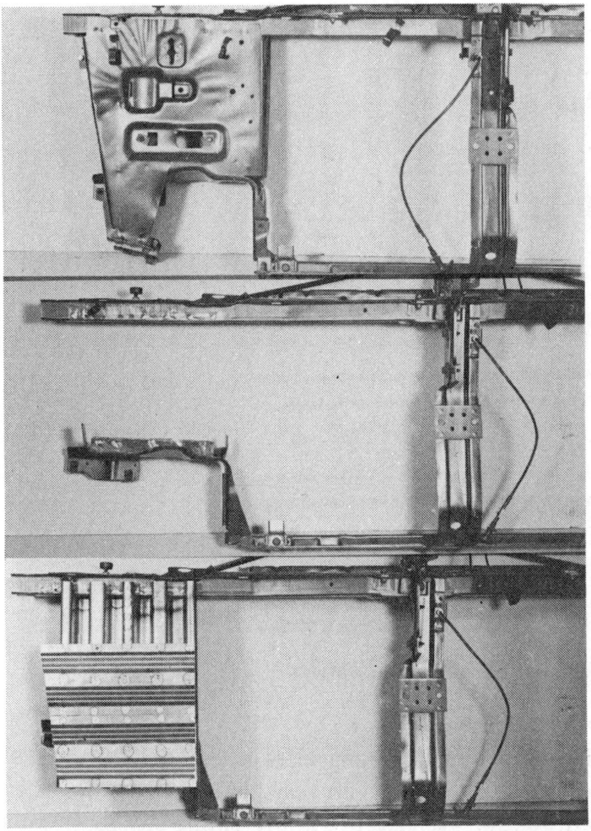

Fig. 4. Original stock radiator support structure (on the top) and rebuilt performance (at the bottom).

maintained without change. Because of the increased stroke of the new system, the headlamp support structure and the hood latch support member were also revised to provide required impact clearance. The headlights were maintained in the stock position. The stock grilles, made of rigid ABS, were removed during impact testing. In production, the grilles can be molded integral with the flexible fascia or they can be separately molded of a flexible material. The parking lamps were mounted in the reinforcement (flex straddle) that was molded into the energy absorber and were recessed slightly more than in the stock vehicle. These lamps were protected in the impact by the reinforcement and during the impact moved rearward with the reinforcement.

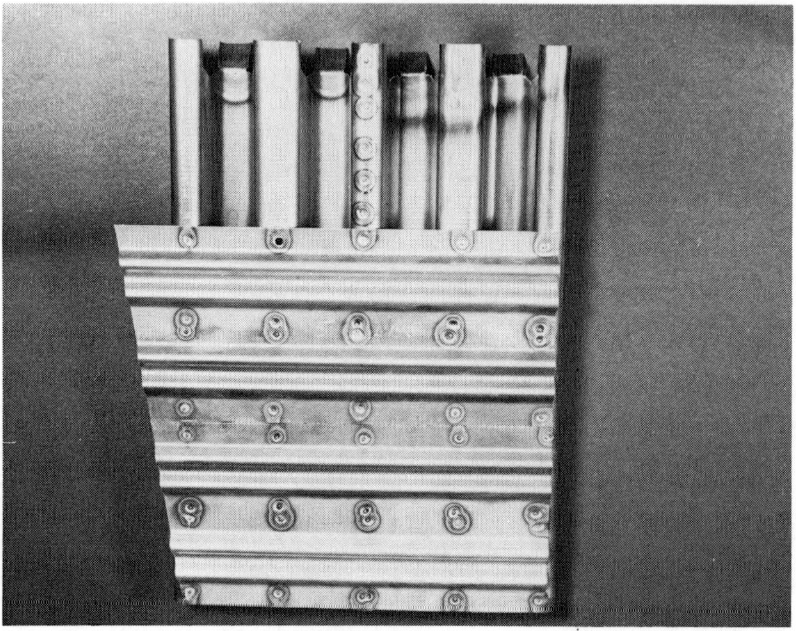

Fig. 5. Stock radiator support from Martensit steel panels.

This vehicle was then subjected to a rigorous series of tests to demonstrate that it not only more than met all the requirements of Phase II of the U.S. Bumper Standard Part 58, but was effective in preventing damage in a variety of real world impacts. Whereas certification testing for the U.S. Standard requires only four pendulum impacts and one barrier impact, nine pendulum impacts covering worst case conditions, and one barrier impact were run on the same set of components with zero damage, denting, or set. Table 4 gives the forces that were experienced during this series of tests. Note that the total barrier force on this heavy vehicle (1 800 Kg) was only 58 KN. This compares with 109 KN for the stock 1978 vehicle. Thus, the forces have been reduced to a very modest "g" loading of 3.3 which, if applied to a 1 000 Kg vehicle, would mean a total barrier force of only about 32 KN or rail loadings of only 16 KN.

Testing was then extended to include a series of real world impacts (still with the same single set of components). Table 5 lists these tests and the results. In all but one case, the Firebird was the impacting vehicle. All

Table 4. Firebird deep foam demonstrator.

Impact forces, KN
Vehicle weight – 1 800 kg

-BARRIER	58 KN
-CENTER PENDULUM	22 KN
-OUTBOARD PENDULUM	29 KN
-CORNER PENDULUM	16 KN
-PLANE A MAXIMUM	33 KN
-MAXIMUM "G" LOADING = 3.3	

Firebird tests, except No. 7, were run with a single production fascia. Test No. 7 was run as an afterhought after the original stock Trans Am fascia had been put back on the vehicle. The energy absorber was not chaged. Because the Aspen bumper did not stroke, the energy absorbed by the Firebird demonstration system in test No. 7 was equivalent to a 7 mph impact into a fixed barrier. No measurable or visible damage was incurred by the Firebird throughout the entire series of tests. The purpose of Tests No. 3 and 4 was to confirm the reduced hostility of the deep foam bumper compared to

Table 5. Real world impact test series front firebird deep foam bumper.

	TEST SPEED	
1. CENTER OF FIREBIRD INTO FRONT CORNER OF 1978 ASPEN BUMPER AT 30°	3.5 MPH	-NO DAMAGE.*)
2. HEAD-ON INTO FRONT OF 1978 ASPEN.	5 MPH	-NO DAMAGE.
3. FRONT OF FIREBIRD INTO RIGHT SIDE OF 1978 ASPEN AT 90° CENTERED AT PILLAR POST.	6 MPH	-NO DAMAGE TO FIREBIRD. ASPEN FULLY OPERABLE. $403. COSMETIC DAMAGE.
4. FRONT OF 1978 ASPEN INTO LEFT SIDE OF 1978 ASPEN AT 90° CENTERED AT PILLAR POST.	6 MPH	-NO DAMAGE TO IMPACTING ASPEN. DOORS AND WINDOWS NON-FUNCTIONAL. $704. COSMETIC & STRUCTURAL DAMAGE.
5. CENTER OF FIREBIRD INTO 55-GALLON SAND FILLED DRUM	5 MPH	-NO DAMAGE TO F/B°). BARREL TIPPED OVER & SLIGHTLY FLATTENED ON IMPACTED SIDE.
6. CENTER OF FIREBIRD INTO UTILITY POLE.	5 MPH	-NO DAMAGE.
7. HEAD-ON INTO FRONT OF 1978 ASPEN	10 MPH	-NO DAMAGE TO FIREBIRD. TWO SMALL GRILLE BARS CRACKED ON ASPEN.

*) Where "no damage" is listed, it includes both the firebird and the impact item.

°) As the barrel tipped over, the corner caught the lower edge of the fascia stretching it out severely, but still with no damage.

Table 6. Weight comparison 1977 – 78 firebird & deep foam demonstrator complete front end.

	1977-78	DEEP FOAM DEMONSTRATOR
-FULL FASCIA WITH GRILLES	5.7	5.7
-ENERGY ABSORBER (FOAM)	2.3	5.7
-BUMPER BEAM	17.3	
-FLEX STRADDLE		5.2
-LAMP & FASCIA SUPPORT (Including Headlights)	6.8	7.0
-HANGER BRACKETS (2)	7.7	
-RADIATOR SUPPORT ADDITION		1.8
-MISCELLANEOUS HARDWARE	1.6	1.6
	41.4 Kg	27.0 Kg

a metal stroking bumper. Damage to the side of the Aspen in the deep foam impact was purely cosmetic with no structural damage or loss of function whatsoever. In contrast, the other side of the Aspen impacted by the steel stroking bumper suffered severe structural damage with complete loss of function of both doors and windows.

This system is also very light, as shown in Table 6, with a weight savings of 14.4 Kg compared to the stock vehicle. Notice that these weights include the upper fascia, grilles, headlamp supports and even the headlamps. Since most bumpers do not include anything above the bumper itself, Table 7 shows the weight and estimated cost of the deep foam bumper only so that it can be compared directly with other bumper systems. In the first line, the

Table 7. Deep foam bumper*) weight and cost estimates.

	WEIGHT % OF VEHICLE WEIGHT	COST % OF VEHICLE COST
-CONVERSION OF EXISTING VEHICLE. 8 KPH U.S. STANDARD 581	0.85	1.4
-VEHICLE DESIGNED FOR OPTIMUM. 8 KPH U.S. STANDARD 581	0.5	1.0
-VEHICLE DESIGNED FOR OPTIMUM. ISO 4 KPH STANDARD	0.3	0.6

*) Bumper only. Blade fascia, E/A, Flex Straddle, Attachment.

weight and cost of the demonstrator is shown as a percent of vehicle weight and cost, respectively. These values can be considered as typical of a conversion of an existing vehicle to a deep foam system that will meet all the 5 mph requirements of Phase II of U.S. Bumper Standard Part 581. In the second line is shown what can be achieved when the vehicle is designed from the very beginning for this type of bumper, again meeting the full 5 mph requirements of the U.S. Standard. Conventional bumpers meeting this standard are typically more than twices as heavy and are more costly. This comparison does not take into account the reduction in structure behind the bumper or other secondary weight reductions.

In the bottom line is shown the weight and cost of an optimum deep foam bumper system that will meet the requirements of the proposed ISO 4 KPH bumper standard. The fact that many EMP bumper face bars are being used on European cars was very apparent at K 79. European manufacturers are challenged to compare these and other bumper systems currently in use in the European market with the values shown in the bottom line of Table 7. In making the comparison, the bumper face bar, any reinforcements behind the face bar, the attachment brackets and any sight shields, corner fillers, and rub strips should be included. It is anticipated that such a comparison will show that this new generation beamless deep foam system will provide a reduction in weight at a surprisingly competitive cost, particularly if advantage is taken of the low forces by reducing the backup structure as well. In addition to these advantages, the vehicle owner can be offered a bumper that fully meets all the tests required by the proposed ISO 4 KPH Standard as well as many real world impacts with no damage to either the vehicle or the bumper. Beyond this, these bumpers offer the stylist the opportunity to integrate the bumper cover smoothly into the body lines of the vehicle for improved aerodynamics. Furthermore, the finish is not limited to dull colors that contrast with the body finish but can be made to match the body in color and gloss.

Pedestrian protection

An added benefit that should be particularly important in Europe and Japan is that the softness of this bumper will provide a significant degree of pedestrian protection. The real world tests referenced above demonstrate that this new generation of bumpers is less hostile to anything it may strike. That can include the pedestrian. In the United States, the current pendulum requirements of Standard Part 581 prevents achieving the optimum in pedestrian protection. In Europe, there is no such barrier to good bumper design. Without the impediment of an arbitrary standard, automobile manufacturers in Europe and the other nations of the world have the potential of utilizing the weight, cost, damage and pedestrian benefits of soft bumpers even more effectively than is possible in the States.

In spite of the limitations imposed in the States, and even without the addition of pedestrian protection requirements, it is expected that the use of soft bumpers will expand to fifty percent or more of the U.S. production by 1985.

Exterior rigid body components

The next question is, where else can RIM polyurethane be used to advantage? Beyond flexible fascia, the next major area of application appears to be in rigid parts to replace exterior body components.

By combining the reformulation of the urethane to obtain high rigidity with the addition of fiber reinforcement, new materials have been developed which have, a) the high rigidity required to replace steel in exterior body panels, b) the reduced thermal coefficient of expansion required for compatibility with mating steel structures, and c) improved high temperature

Table 8. Typical properties – Davidson rubber RRIM formulation thickness 100" (2.5 mm).

SPECIFIC GRAVITY ASTM D792	1.2
TENSILE STRENGTH, PSI (MPa)	5000 (30,000)
ELONGATION, % ASTM D638	25
FLEXURAL MODULUS, PSI (MPa) ASTM D780 @ 75°F (24°C) @-20°F (-29°C) @158°F (70°C)	 250,000 (1,700) 450,000 (3,100) 145,000 (1,000)
HEAT DEFLECTION TEMPERATURE °F (°C) ASTM D648	250 (120) @ 0.150" (3.8 mm)
IZOD IMPACT, FT.-LB/IN (J/CM) ASTM D256 @ 75°F (24°C) @-20°F (-29°C)	 2.6 (1.4) 1.5 (0.8)
THERMAL COEFFICIENT OF EXPANSION ASTM D696, IN/IN/°F (°C) x 10^{-5}	3 (5)
SHRINKAGE, %	0.5

*) Properties are the average of two directions, parallel and perpendicular to flow.

resistance for better paintability. The properties of one of these materials are shown in Table 8. Components that can be considered as potential applications for this material and others like it are as follows:

Fenders
Rocker Panels
Door Outer Panels
Grille Opening Panels (Header Panels)
Hood Outer Panels
Deck Lid Outer Panels

Both Ford and General Motors have announced their intentions to commercialize rigid reinforced RIM (RRIM) fenders. Production trials are expected to take place in 1981-82 and production programs are expected for 1982-83.

The basic reasons for concentrating on fenders as the first application are as follows:

1. *Weight Savings.* It is estimated that a RRIM fender will be half the weight of a steel fender at about the same cost.
2. *Limited Structural Requirements.* Fenders are largely cosmetic and are required only to hold themselves in position with no external applied loads.
3. *Integration of Pieces Into One Component.* The very high moldability of RRIM makes it possible to combine what would be a welded or bonded assembly of two or more pieces in steel or SMC into a one piece molding for additional weight and cost savings.

In addition to the distinct and primary weight and cost advantages provided by its low specific gravity and high moldability, RRIM is favored by many of the same advantages as flexible RIM, such as:

1. *Excellent Paintability.* Compared to other plastics, it has an excellent durable, glass-free surface such that the color, gloss, and DOI can be made to match adjacent metal surface finishes.
2. *Damage Resistance.* Note in Table 8 that the elongation is many times that of SMC, giving it both in-plant and in-field damage resistance. Because of this quality, it has been called "the friendly fender".
3. *Corrosion Resistance.* These materials have excellent weathering characteristics and eliminate problems associated with galvanic action. They do not rust.

4. *Low Tooling and Retooling Costs.* Because internal mold pressure is low (less than 700 KPa), molds need not be built to withstand the high molding and clamping of injection and compression molding (10 to 20 MPa).

5. *Low Energy Consumption.* The energy required to produce parts in these materials is lower than steel or aluminium, even including the feedstock from which the material is derived. At the point of scrapping, the heating value of the material as fuel remains to be used.

A prim application case study

To prove feasibility and performance, Davidson Rubber Division of Ex-Cell-O Corporation has prototyped a RRIM fender as an international development program. To do so, an Omni four-door sedan was purchased as representative of both European car design and the new generation of U.S. front wheel drive vehicles. The steel left front fender was removed and used

Fig. 6. RRIM fender with high quality surface.

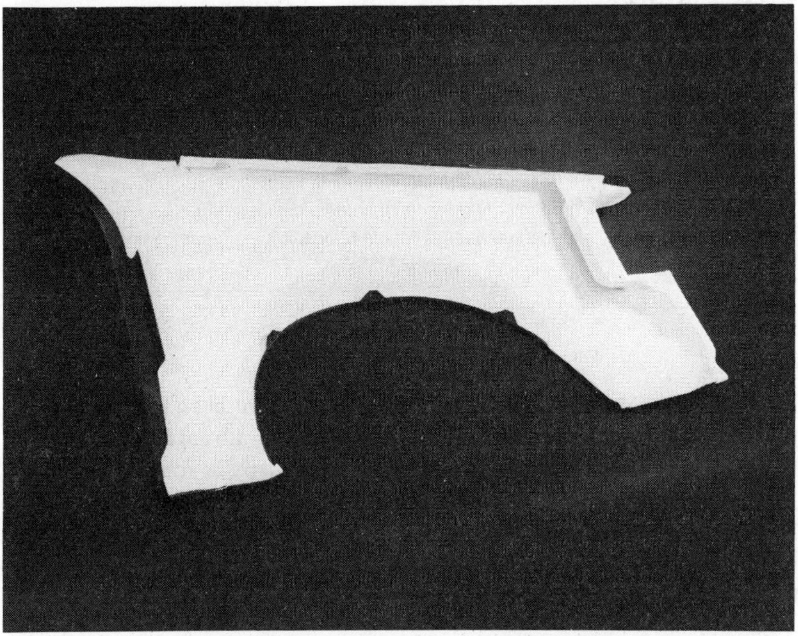

Fig. 7. Back side of a RRIM fender with integration of components.

as the model for making a mold for a direct replacement RRIM fender. The resultant molded RRIM fender shown in Figure 6 illustrates the high quality class A surface attainable, fully comparable to the sheet metal it replaces. Figure 7 is a view of the back side and illustrates the moldability of RRIM and the potential for integration of components. In case of the steel fender, the park and side marker lamp housing was separately stamped and welded to the fender. As shown in Figure 7, it was molded as an integral part of the one piece RRIM molding. This configuration would be impossible to mold in one piece in SMC for instance. Demonstrating effective weight saving, the RRIM fender weighed 2.1 kg's compared to 4.2 kg for the steel fender.

To demonstrate the real world in-field damage resistance of the ''friendly RRIM fender", a series of tests were run comparing the RRIM fender to the steel fender. Table 9 lists the tests and the results. Note that in each test, there was significant damage to the steel fender whereas the RRIM fender suffered little or no damage.

Table 9. RRIM fender VS steel fender damageability tests

TEST	DESCRIPTION	STEEL FENDER	RRIM FENDER
STONE IMPACT	WALNUT SIZED STONE FIRED FROM SLING SHOT	1/2" DEEP DIMPLE	NONE (SCUFF MARK)
SHOPPING CART	LOAD - 60# VELOCITY - 8 KPH	3/4" DEEP DENT	NONE
ADJACENT CAR DOOR	VELOCITY 6.5 KPH	1/2" DEEP DENT	NONE
GARAGE DOOR	1 1/2" INTERFERENCE	SEVERE DAMAGE, REPAIR COST = $135.	SCUFF MARKS AND SCRATCHES. TOUCH UP PAINT = $2.

In addition to the Omni, several other fenders have been prototyped, further verifying the feasibility, practicality, and advantages of RRIM fenders. In some cases, a welded assembly of as many as four separately stamped components has been replaced by one relatively simple RRIM molding. Our next project will be the molding of an outer door panel, which will be adhesively bonded to the inner panel and frame structure, providing both weight reduction and damage resistance.

Summary

A rapidly growing population of flexible fascia has established a firm position for RIM polyurethane in the U.S. automotive market and a similar giant may well awaken in Europe and other places.

Now the extension of urethane chemistry into rigid RRIM body components gives promise of an even greater worldwide market. While great strides have been made in the development of RRIM materials and components, much work remains to be done. Production scale processing and long term field performance remain to be determined. However, there is a worldwide need for weight reduction, and it is a fact that a very high proportion of car damage claims involve a fender or door panel. These are important factors in making RRIM particularly attractive for exterior automotive body components.

Bibliography

[1] *Ferrari, R.J.:* "Rigid RIM – A microcellular Structural Foam", Paper No. 760334, SAE International Automotive Congress & Exposition, Detroit, MI, 1976.

[2] *Weller, P.A.:* ''Weight Savings in Automotive Body Components Using Plastic'', Technical Report No. 76-18, American Society for Metals, Metals Park, Ohio, October 1976.

[3] *Weller, P.A.,* and *J. V. Scrivo:* ''Davisorb Bumpers Reduce Weight & Damage'', Paper No. 770308, SAE International Automotive Congress & Exposition, Detroit, MI, 1977.

[4] *Scrivo, J. V.:* ''1979 Firebird – An Advanced Soft Bumper System'', Paper No. 790335, SAE International Automotive Congress & Exposition, Detroit, MI, 1979.

Authors

Basille	Michel, Dipl.-Ing. Technischer Direktor ste. Quillery · F-92250 La Garenne – Colombes Tel. (1) 7807302 · FS Quiller 620875 – F
Bertone	Nuccio President of Carrozzeria Bertone S.p.A. Corso Allamano 40/46 · I-10095 Grugliasco – Torino Tel. (011) 788044 oder 783222 · FS 220143
Fernengel	Rudolf, Ing. VDI Bayerische Motoren-Werke AG BMW-Kunststoffentwicklung Postfach · D-8000 München 40 Tel. (089) 3183204/278 · FS 528248
Fike	W.H., Chief Eng. Ford Motor Company Plastic Development and Appl. Office 24300 Glendale · Detroit Michigan 48239/USA Tel. (313) 5922423 · FS 235504
Fischer	Wolfgang, Ing. grad. Daimler Benz AG, Abt. A 1 RS Postfach 226 · D-7032 Sindelfingen Tel. (07031) 6122795 · FS 7265721
Glagow	Klaus C., Executive Eng. Vehicle Engineering Dep. Ford of Australia Ltd. Box 14 Geelong · Victoria 3220 Melbourne / Australia Tel. (0061) 52780211 · FS Fords AA 30762

Hablitzel	Hermann, Ing. VDI Volkswagenwerk AG Leiter der PKW-Entwicklung D-3180 Wolfsburg Tel. (05361) 223570 · FS 9586533 vwwd
Heim	Reinhold, Ing. Daimler Benz AG, Abt. A 1 AM Postfach 226 · D-7032 Sindelfingen Tel. (07031) 6122389 · FS 7265721
Hellriegel	Edmund, Ing. Ford-Werke AG, Abt. MC PEF-R Postfach 604002 · D-5000 Köln 60 Tel. (0221) 7197439 · FS 888483
Iseler,	Kenneth A. Director of Research and Development for the Plastic Division The Budd Company 3155 West Big Beaver Road Box 2601 · Troy, Michigan 48084/USA Tel. (313) 6433723 · FS 0023-0235-837
Johnke	Klaus Dieter, Ing. VDI Volkswagenwerk AG PKW-Entwicklung D-3180 Wolfsburg Tel. (05361) 224821 · FS 9586533 vwwd
Klie	Wolfgang, Dipl.-Ing. Daimler Benz AG, Abt. A 1 GS Postfach 226 · D-7032 Sindelfingen Tel. (07031) 6122795 · FS 7265721
Noris	Claudio, Dr. Stars, Division of Comind S.p.A. Corso Savona 45 · I-10029 Villastellone (Torino) Tel. (011) 9698683 · FS 210305

Quinn	James, N. Group Vice President Plastic Products The Budd Company 3155 West Big Beaver Road Box 2601 · Troy, Michigan 48084/USA Tel. (3 13) 6433723 · FS 0023-0235-837
Röpke	Reinhard, Dipl.-Ing. Geschäftsführender Gesellschafter Westfälische Metall Industrie KG Hueck & Co. Postfach 2840 · D-4780 Lippstadt Tel. (02941) 387346 · FS 84941
Stübig	Hermann, Ing. (grad.) Audi NSU Auto Union AG Gesamtplanung D-8070 Ingolstadt Tel. (0841) 892289 · FS 55801
Travis	D.C., General Manager Ford Motor Comp. USA Plastics, Point and Vinyl Div. 24300 Glendale · Detroit Michigan 48 239/USA Tel. (313) 5922423 · FS 235 504
Weller	Peter A., Director, Product Engineering Davidson Rubber Division Ex-Cell-O Corporation Industrial Park · Dover, New Hampshire 03820 Tel. (603) 742-0720
West	Michael W., Eng. Application Development Specialist General Electric Plastics Auto Polymer Center Telegraf Road, 25900 · South Field, Michigan 48034/USA Tel. (3 13) 5523200

Zumhagen Conrad F., Eng.
Programs and Development Manager
General Electric Plastics
Auto Polymer Center
Telegraf Road, 25900 · South Field, Michigan 48034/USA
Tel. (313) 5523200